International Energy Futures
Petroleum Prices, Power, and Payments

Nazli Choucri

with David Scott Ross
and the collaboration of Brian Pollins

The MIT Press
Cambridge, Massachusetts
London, England

© 1981 by
The Massachusetts Institute of Technology

This book was set in Fototronic Times Roman by The Colonial Cooperative Press Inc. and printed and bound by The Alpine Press, Inc. in the United States of America.

Library of Congress Cataloging in Publication Data

Choucri, Nazli.
International energy futures.

Bibliography: p.
Includes index.
1. Petroleum industry and trade—Finance—Mathematical models. 2. Petroleum products—Prices—Mathematical models. 3. Energy policy—Mathematical models. I. Ross, David Scott, joint author. II. Pollins, Brian, joint author. III. Title.
HD9560.6.C49   338.2'3   80-39890
ISBN 0-262-03075-6

International Energy Futures

Contents

Preface

This book is the result of many years of work and the contribution of many individuals. The study began in 1970 as a pilot analysis and continued from 1972 to 1975 with support from the National Science Foundation. A report to the foundation summarized the results to date. In the next two years the interconnections between political and economic factors in the world energy market were more fully specified. The initial report was the background document for this reassessment. The conceptual basis for analyzing interdependence in resource markets was developed with assistance from the Rockefeller Foundation. While there is every prospect that specific aspects of an analysis of the international energy and petroleum situation can become readily out of date, the essential characteristics of the oil problem and the interconnections between economic and political factors will undoubtedly persist for decades to come. The book was completely updated in the last two months of 1979.

This book is organized in four parts. Part I presents the energy problem from the different perspectives of the producing countries, the oil-importing countries, and the international oil companies. A general model for analyzing interdependence and forecasting future relations is described. Part II examines the structure of the model in detail. It is mainly for the reader who has a technical background. Part III presents the basic results and compares these with other existing evidence. Part IV provides an analysis and forecast of alternative, hypothetical outcomes. The policies of the oil-exporting countries are examined as are some responses by the oil-importing countries and the international oil companies. It is a study of futures: What will happen if . . . ?

While I alone can be held responsible for the contents, arguments, and results, I gratefully acknowledge the assistance of colleagues at MIT and elsewhere who contributed to different aspects of the analysis. M. A. Adelman helped me to understand the oil production process and appreciate the intricacies of the industry internationally. Dennis Meadows, who collaborated in the first phase of the study, made important suggestions at the start. Davis Bobrow reviewed an early version of the book. Hayward Alker, Lance Taylor, and other colleagues made valuable suggestions.

To the research assistants of this project I owe the greatest debt of all. Tariq Panni, then a graduate student in the Sloan School of Management, contributed to the conceptual and computational foundations of the project during its first phase. Subsequently David Scott Ross, then in the doctoral program in the Department of Political Science, wrote the second version of the computer program for the model that appears, with major revision, in part II of this book. Brian Pollins, who joined the project after its inception, contributed valuable research assistance in debugging and to a considerable extent in tracking down conceptual difficulties and attendant computational problems. Zale Anis provided computational assistance as well. Peter Brecke and Dale Smith contributed to organization and design. The final phase of the analysis entailed projections of alternative futures, now parts III and IV of this book. Michael Lynch, formerly a graduate student, and presently on the staff of the Energy Laboratory, worked on the validation, design, and analysis of these scenarios. Many other individuals too numerous to identify here also contributed numbers, assessments, or simply opinions to assist in the documentation of this book. I am grateful to all.

Nazli Choucri

I
The Problem

1
International Energy Policy

1.1 The Politics of Petroleum Trade

The increase in petroleum prices of October 1973 was economic in nature, buts its effects were political as well. The impacts were far-ranging, although no one fully understood their ramifications. The dominant view in the West was one of "crisis" and the price rise almost uniformly regarded as detrimental to the interests of the oil-importing countries. Various opinions were expressed as to whether the crisis was artificial, resulting from contrived changes in market mechanisms, or whether it was real, created largely by actual shortages in petroleum supplies.

Price determination in the world oil market has never been entirely governed by the mechanisms of a competitive market. The cost of petroleum production in the Middle East has always been far below price, even before the 1973 increases, and even then supply and demand did not govern world prices. Almost everyone agrees that a wide margin still remains for further price increases. In the long run an upper-price threshold will be set by the commercial availability of alternative energy sources and in the shorter term by the degree of cohesion among the oil-exporting countries and the willingness of suppliers to meet demand. The policies of the oil-exporting countries will be instrumental in shaping both short- and long-term conditions.

The price of oil can be viewed simply as a rate of payment to owners of productive resources. But price is also a reflection of power: it reflects the control of countries and companies over rates of payment. Both the politics and economics of oil prices impose a strong interdependence among buyers and sellers. Until the events of October 1973 this interdependence in the world petroleum market was not fully appreciated. The initial price rises drew attention to mutual constraints and to the fact that the sellers' policies created macroeconomic as well as microeconomic resource-adjustment problems for the buyers.

The interdependence between buyers and sellers highlights still more basic political conflicts over who controls price, who determines output, and how international trade in petroleum is regulated. Today's major political debates range principally around pricing, production, and trade policies. The new exercise of political power by the oil-exporting countries has created problems of access and security of supplies for the oil-consuming countries, the reality of which is difficult to ignore.

Responsibility for today's oil prices has been attributed to the exporting countries for raising petroleum prices in a seemingly arbitrary manner, to the importing countries for placing extensive demands upon existing petroleum reserves, or to the multinational corporations for allegedly encouraging the producers in their price escalation policies. The price increases are defined differently depending on the interests, goals, and objectives of the parties involved. But the underlying cause is clear: there is new interdependence among buyers and sellers in the world petroleum system; there are competing interests, goals, perspectives, explanations of "crisis," and prescriptions for resolution. While the oil exporters are becoming more powerful, and the oil importers are searching for "solutions," the oil companies continue to play an important role in resource allocations internationally.

It is not at all clear, however, where effective control over economic policies lies: both oil exporters and importers appear vulnerable to the actions of each other, with varying financial and political consequences. There is no "average" oil-importing or oil-exporting country: aggregation is rather a prescription for simplification; when the energy issues are viewed from different vantage points, they appear, and in fact are, very different.

The dependence of consumer economies on petroleum is unique: no other natural resource has the effect or influence on economic activity as oil. Between now and the turn of the century the difference between demand and the availability of conventional energy sources, notably oil, defines the nature of the energy problem for the consumer countries of the West. This problem includes

1. curtailing demand for energy,
2. assuring access to and availability of petroleum,
3. financing payments for the oil imports,
4. developing alternative sources of petroleum and domestic reserves,
5. developing and investing in alternative energy sources,

6. confronting a situation of strategic vulnerability due to oil import dependence, higher oil import payments, insecurity of supplies, and the continued importance of oil in industrial processes.

The price increase of October 1973 was not an isolated incident. The oil-exporting countries had tried in 1967, with little success, to raise the price of crude oil, partly in protest against the Arab-Israeli war, but more basically in reaction to changes in the world market and perceived inequities in prevailing prices. In October 1973 a series of circumstances converged to enable the successful price increases, production cuts, and embargo threats. Changes in the world oil market and in the relationship between the international oil companies and the governments of the oil-exporting countries contributed to their success in 1973. In addition the changes in regional politics in the Middle East and the development of a more pragmatic stance by both rich and poor states alike enabled the price of oil to be increased as an instrument of foreign policy. Economic issues became intensely politicized.[1]

The oil-exporting countries have argued that the historical oil prices did not reflect relative scarcities in the oil market. Today their objectives are to generate a "just" price of crude oil and participate in the development of alternative energy sources. But between now and such time as the role of petroleum in industrial processes is significantly reduced, the oil-exporting countries will face the following problems:

1. absorbing and disbursing their oil revenues,
2. making decisions regarding alternative types of economic investment,
3. coordinating production policies,
4. setting tax rates and/or production schedules to protect the value of their assets,
5. conserving oil reserves,
6. managing the economic interdependence with consumer countries, given their own foreign investments, their requirements for advanced technologies, and their susceptibility to pressures for increased production.

The degree of cohesion among the oil-exporting countries will undoubtedly influence these issues. The First Arab Energy Conference of March 1979 resulted in a new orientation for many of the oil-exporting countries: a greater concern for rationalizing their production schedules, an insistence that the policy debates revolve around the price of final products to consumers, and a shift toward greater emphasis on alternative sources of energy. The 1979 revolution in Iran and the cutbacks in production clearly affected the oil market, providing the urgent stimulus for adjustments by the consumer countries.

The international oil companies continue to play a critical role in the world oil market. Historically the operating companies in the oil-exporting countries produced and transferred oil at a fraction of the actual costs to the shareholder companies or affiliates. Prior to the events of October 1973 this oil was exchanged at posted prices to other affiliates, generating profits, a portion of which was paid as tax to the host government. But the 1973 increases in producer government revenues did not reduce the net profits to the oil companies; the companies just passed the increases on to the consumer.

Today, with the escalating prices of oil the major oil companies are faced with certain imponderables. There are no viable alternatives to Middle East oil in the next decade. The growth of independent petroleum companies—and more recently of national oil companies—constrains the flexibility of the majors in pricing their goods and services. With projected growth in the petroleum imports of the United States and other consumer countries, the pressures placed upon Middle East sources, in particular Gulf oil, will increase.

Among the traditional concerns of the oil companies are the following:

1. generating profits from oil sales,
2. determining appropriate markup due to changing supply and demand relationships,
3. undertaking exploration and development,
4. investing in the oil industry,
5. allocating market shares internationally.

Changes in international power relations affecting the world oil market will invariably influence the role of companies further. National oil companies of producer countries increasingly provide alternatives to the multinationals in undertaking certain functions for the oil industry. The development of new sources of petroleum will place new constraints on both the companies and the oil-exporting countries and provide new leverages for the consumer countries.

As control over oil trade and energy developments are being disputed, power relationships are being changed. The governments of the oil-producing countries are now dominant agents in the world oil market. The criticality of oil production to the economies of the importing countries and the absence of competitive substitutes in the short run contribute to the reinforcement of a price-setting mechanism by which the sellers essentially control price. In this market price is set less by the competitive relationship of supply and demand—although market conditions impose some constraints on the sellers—than by the degree of cohesion among the suppliers. Today the Organization of Petroleum Exporting Countries sets the prices, subject to certain, potentially important constraints.

Consequently policy concerns for importer and exporter countries cannot be addressed within the context of an economic analysis that pays insufficient attention to the prevalence of dual types of influences on oil-pricing policy—the country-based (OPEC) influences and the persistent influence of the international oil companies in allocating market shares and regulating production schedules. If the decisions and interactions among buyers and sellers were governed exclusively, or even largely, by economic motives, then a conventional approach to the analysis of future oil prices would be appropriate. The conventional economic paradigm of supply and demand adjustments to price would be the best method of analysis. But the oil problem is not one that can be understood on narrow economic grounds, and the demands governing the oil industry world-wide have been influenced by political interventions and political activities of both buyers and sellers. Therefore an analytical framework that recognizes the interconnection of economic and political factors in the world oil market is appropriate. Further a view that differentiates among producer countries, consumer countries, and international oil companies would pose a greater analytical and empirical challenge than one that ignores this difference.

1.2 An Implicit Theory of Price Determination

We propose a new, more realistic view of international oil trade which we have embedded in the International Petroleum Exchange (IPE) simulation model. The IPE model specifies the worldwide effects of different rates of payment per barrel of oil. It is not designed to give a view of an economic "market" as it is conventionally defined but rather to represent an "exchange" of interests and the interdependence of the oil-exporting countries, oil importers, and international oil companies. The exchanges are described in terms of supply and demand relationships, competitive and noncompetitive characteristics of the oil market, trade in oil, the impact of substitute energies, corporate profits from oil sales and investments in the oil industry, balance of payments and national security for consumer countries, and the development allocations and investments of the producer countries. The IPE model was developed at MIT, with the initial support from the National Science Foundation. An early version was completed at the Center for International Studies, and the later versions in the Department of Political Science.

Price as well as supply is influenced by broad political factors, and the adjustments to price of quantities demanded and amounts supplied are also influenced by political factors. The tax rate per barrel of oil has in the past been set by the oil companies in negotiation with the producer governments. In October 1973 the producer countries successfully exerted their influence and together set the tax rate at an unprecedented level. This was an economic move, but with profound political and economic implications. While the exporter countries' oil production policies can be understood on economic grounds, in terms of producing that amount that will help sustain the desired price, it is also a political move. It demonstrates the exercise of sovereignty over national resources and effective control over available supplies.

Since in the petroleum market competition among buyers and sellers is not the mechanism that moves the price and quantities demanded and supplied toward equilibrium, the role of supply and demand in price setting is less determining than in a competitive market. The implicit bargaining process between buyers and sellers in this type of market does not lead to the same market-clearing dynamics as in a competitive situation. By acknowledging the dominance of noncompetitive characteristics, we adopt a view of price determination that considers both the effects of oil-producing countries setting the tax rate and the market-related influences on price determination.

The theory of price determination embedded in the IPE model stresses the interdependence of all three entities: the producer countries, the consumer nations, and the international oil companies. Since its inception, the oil industry has been regulated and controlled by the oil companies. The emergence of direct interaction between importers and exporters of crude petroleum is a relatively new phenomenon that represents a change in the world oil market but does not exclude the continuing role of the international oil companies.

The producer countries set the tax rate on the barrel of crude oil. They affect the balance of payments situation of the consumer countries, not only by the magnitude of the tax rate but by the volume of imports of goods and services from advanced industrial countries. While the essential economic factors of a dramatically increasing demand for oil, depleting a commodity that has no commercially viable substitutes, contribute to collusion among the producer countries, the prospect of political gain has served to consolidate their cooperation and led directly to their exploitation of economic advantage. A perception of injustice in the prevailing price structure has strengthened the political cohesion of the producer countries. As a result the oil-exporting countries chose to influence the market not only through their tax rate policy but through production control schedules by alternatively accelerating, maintaining, or cutting production.

The consumer nations that import petroleum may in turn exert an influence on the terms and nature of exchange by regulating their imports from the Gulf region, controlling demand, expanding their domestic production, or investing in substitute energies. What they do will influence the final price. Consumer decisions can have reverberating effects throughout the relationships of interdependence and exchange.

Though not legally the owners of the resource the international oil corporations have historically undertaken managerial functions that have been central to the oil industry. They exerted a powerful control over all aspects of the oil industry. Today, however, this influence is markedly reduced. In fact the specific influence of the companies on price may appear today to be marginal. But this view is misleading. In response to the conditions in the industry, the companies make oil investment decisions that can influence price. In addition the repatriation of corporate profits will influence the consumer countries' balance of payments.

This formulation of the price-setting mechanisms stresses noncompetitive price setting while at the same time makes provision for the influence of competitive relationships. The price specifications in the IPE model provide a more comprehensive perspective on the influence that buyers, sellers, and oil companies exert upon trade and the terms of exchange. The essential feature of this formulation is that price is generated by the model as the outcome of the policies of the producers and their interactions with consumer nations, international oil companies, and the interdependence that results from these interactions. The final price is not simply the outcome of the producers' policies but of overall interactions among the agents in the market. Figure 1.1 presents the historical oil prices from 1970 to 1980.

1.3 Costs and Benefits

Countries do not have one goal, one threat, one perspective, but many. To evaluate the threat to consumer countries of higher oil prices, or to producer countries of lower oil revenues, or to oil companies of lower profits, one must recognize the diverse views of the problem and the differences in the perspectives of each entity in the world oil market. What is re

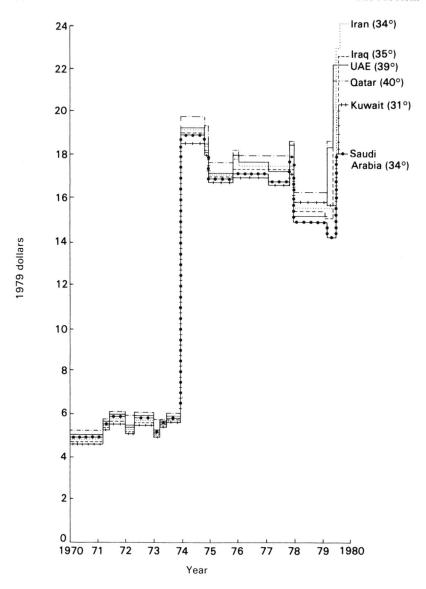

Figure 1.1 Oil prices

garded as a threat to the oil-importing countries may be viewed with favor by the oil-exporting countries. Perceptions of threat are intensely subjective, but they may assume a strength that transcends their empirical reality. Perceptions of threat are often more important in determining national responses than are the realities of a situation. Alternative tax rates— and their affect on oil prices—represent real and perceived threats to producers, consumers, and companies. What might appear as an economic threat also has political ramifications. Political actions to be effective must consider economic consequences.

Embedded in economic calculations are political costs and benefits. Strategic vulnerability, for instance, which can be measured by economic indicators, must be interpreted in political terms. Similarly the volume and value of oil revenues for the producer countries can have extensive political implications: if the consumer countries had markedly lower oil prices, for example, if there were no OPEC, the effects would not be uniformly positive over the long run. Lower prices entail costs to the consumers, some of which may not be intuitively obvious. An optimal price for the producer countries based on one criterion may not be optimal when other criteria are taken into account. The concept of optimality must be viewed in terms of all the attendant costs and benefits implied by a policy.

Short-term benefits may create long-term costs. Immediate costs can hide longer-range benefits. In addition there are trade-offs between economic and political gains. Economic policies designed to have certain immediate impacts may create longer-range effects that could be contradictory, even inconsistent with the initial intents. The path of a policy outcome over time may have peaks and troughs, and at each point the associated political implications may vary accordingly. The feedback effects in dynamic interdependent systems of interaction are not easy to identify. Determining changes due to the adjustment to initial policy interventions is undoubtedly one of the most important contributions of a comprehensive perspective on international energy issues.

There are also concerns of a global nature in contrast to the interests of individual nations. Depletable resources can be viewed as global assets, and the management of those resources as an international trust to be undertaken by all participants in resources exchanges. This view directly contradicts the opinion that natural resources are national endowments, that the individual nations are the legitimate controllers and arbiters of the terms of exchange, and that there can be no international claims on national endowments. We reconcile these views by recognizing that the value of the resources is a national concern, but the volume and pace of exploitation of remaining reserves are a worldwide concern and the final basis for international coordination in resource exploitation, development, and management.

The effects of five alternative price policies based on different tax rates are examined and projected to 2000. These alternatives must be regarded as hypothetical since, in fact, no one is fully informed as to the future policy intent of the oil-exporting countries.[2]

The historical tax rates of 1970 to 1978, in real terms, were used as the reference for examining the results of the model, with the 1978 rates assumed to persist to 2000. Although this reference is already outmoded, it is useful for comparing alternative projections of the future for different producer countries' tax rates. The tax policies compared to the reference case are

• a sharp tax increase policy that raises the tax rate in 1985 by 100 percent over the 1978 tax rate to $26.60 in 1979 dollars, reflecting a *conservative* policy by present standards;

• a tax cut policy in 1985 by 50 percent to $6.65 in 1979 dollars, labeled the *breakup* of OPEC;

• a policy of setting the tax rate corresponding to *optimal* prices for producer countries;[3]

• a policy that adopts a declining rate from 1970 to 1978 in constant terms, then a constant tax throughout the rest of the period, stipulating no collusion among the exporters or *no OPEC;*

• a *radical* tax policy that follows an optimal path to 1975, then increases by 7.2 percent annually to 1985 and by 5 percent to 2000.[4]

This book identifies many threats, several sources of vulnerability, and alternative criteria for evaluating the gains and losses associated with different tax policies. The summaries in tables 1.1 through 1.3 highlight the costs and benefits of alternative policies for the major participants in international energy exchanges. The resource is crude petroleum, but the effects—and price—of alternative energy sources are also taken into account. In each case the policies are presented roughly from the most to the least favorable, based on general, largely short-term considerations. But the assessment of costs and benefits also takes a longer-term view of the alternative futures for producers, consumers, and oil companies. The judgments are summaries only of the different tax policies examined. Alternative policy interventions on the demand and supply sides include of course conservation measures by the consumer countries which affect demand or cutback policies by the producers which affect supply.

If consumers are concerned only with the price of oil imports, then the no-OPEC tax policy is the most favorable (table 1.1). But such a policy inevitably entails a broad range of long-term economic and political costs that cannot be easily eliminated. Strategic vulnerability, insecurity of supply, and a deteriorating payments position due to greater oil imports, though at low prices, are among the most prominent costs. If consumer countries take into account the long-range effects of low oil prices, they must conclude that the real threat lies in low prices, which create greater political problems of dependence on external sources of energy and economic vulnerability to an insecure oil supply. Incentives for curtailing demand are not strong. Low oil prices will erode the basis for an overall energy policy addressed not to prices alone but to the structure of demand, the balance of payments, domestic oil production, strategic vulnerability, and the availability of substitutes. In these terms lower prices are not in the consumer countries' long-term interest.

For producers a price policy regarded as optimal, based on the present discounted value of future revenue streams, is not optimal when other factors are taken into account (table 1.2). The most favorable policy is in fact a new, sharp increase in oil prices and the maintenance of taxes at the new level. This is not an argument for forcing a continued increase in taxes, rather a recognition that, once a dramatic increase is made, it is advantageous to maintain at least the new level and not reduce taxes, as stipulated by a conventional optimal price strategy. With higher prices the producer countries conserve both their resources and the value of remaining reserves. On all dimensions of national concern the radical price policy dominates the other options considered here. Of course to determine the limit of such a policy is not easy. Further a necessary concomitant is the prevalence of years of excess capacity which has in the past forced prices down.

The companies' interests are tied almost exclusively to the volume of oil trade (table 1.3). With lower prices more oil will be imported by the consumer countries, and corporate markup will adjust accordingly. But there are also important costs: investments in the oil industry will necessarily increase, cutting into profits and the repatriation of profits. Most important, depletion will occur sooner, and reserves will be strained.

1.4 Beyond 1985

The comparisons in tables 1.1 through 1.3 are summary observations of what can—and will—happen under different tax policies, with their attendant economic and political conditions. More likely decisions will be

Table 1.1
Costs and benefits of alternative policies for consumer countries

Policy[a]	Costs	Benefits
No-OPEC	Substitute availability is very low. Vulnerability is very high. Resource exhaustion is imminent by the end of the period. Capital inflow is negligible. Demand often exceeds supply.[b] Demand is very high. Supplies are very insecure.	Oil imports and demand are relatively unconstrained. Consumer oil import expenditures are very low. Consumer reserves and production are high late in the period. Current account and basic balance of payments are the most favorable.
Breakup	Domestic production is low. Substitute availability is low. Vulnerability is high. Capital inflow is low. Oil import gap and demand are high late in the period. Supplies are insecure.	Oil import expenditures are low. Current account is favorable.
Optimal[c]	Oil import expenditures are high. Current account is unfavorable.	Substitute availability is high; vulnerability is low. Capital inflow is high. Demand is not often greater than supply. Demand is low. Supplies are relatively secure for most of the period.
Conservative	Oil import expenditures are high. Current account is very unfavorable. High domestic oil production exists.	Substitute availability is high. Vulnerability is low. Capital inflow is very high. Demand is very low. Oil import gap is low. Supplies are not very insecure for most of the period.
Radical	Highest oil import bill. High domestic production. Deteriorating payments position specifically the current account.	Lowest growth in demand; improvement in vulnerability position; greater availability of alternatives. Improved capital account. Reduced strategic vulnerability.

[a] These labels are for identification purposes only and do not presume or preclude a particular market structure or adjustment.
[b] In this volume this difference is referred to as oil import gap. See chapters 2 and 8.
[c] Calculated as optimal by Pindyck (1978). See chapter 12.

Table 1.2
Costs and benefits of alternative policies for producer countries

Policy	Costs	Benefits
Radical	Excess capacity in some years. Extensive investments in consumer economies that may result in "hostage" assets.	Highest value and volume of remaining reserves. High domestic and foreign investments. Extensive growth of investment income. Highest oil revenue.
Conservative 1985	Periods of excess capacity occur. High foreign investments make producers vulnerable to consumers.	Resources are preserved. Oil revenues are very high. Producer capital, investments, and imports are very high. Investment income is very high, surpassing oil revenue by 2000. Decline rate is very low.
Optimal	Demand fluctuates, causing periods of excess capacity.	Oil revenues are very high. Producer capital, investment, and imports are high, as is investment income. Decline rate is low.
Breakup	Reserves are strained. Oil revenues are low. Industrial capital, foreign investments, imports, and investment income are low. Demand drops early in the period. Decline rate is somewhat high late in the period.	Demand remains high and supplies tight late in the period.
No-OPEC	Resource depletion imminent at the end of the period. Oil revenues, industrial capital, and imports are very low. Foreign investments and investment income are negligible. Decline rate is very high.	Demand grows rapidly throughout the period.

Table 1.3
Costs and benefits of alternative policies for oil companies

Policy	Costs	Benefits
No-OPEC	Oil-related investments and costs, particularly exploration investments and discovery costs, are very high. Resource depletion is imminent at the end of the period. Repatriated profits begin to drop sharply late in the period.	Oil trade and corporate profits are very high. Prices are relatively stable. Corporate markup remains high throughout this period.
Breakup	Oil-related investments are high. Prices are unstable.	Oil trade and corporate profits are high. Corporate markup and repatriated profits are high late in the period.
Optimal	Oil trade and profits are low. Prices are unstable. Corporate markup and repatriated profits are generally low.	Oil-related investments and costs are low.
Conservative	Oil trade is low. Profits are very low. Prices are unstable. Corporate markup and repatriated profits are low late in the period.	Oil-related costs and investments are low.
Radical	Corporate markup is a low proportion of price. Investments are not strong. Profits are low due to low volume of trade.	Reserves are not depleted. Costs do not grow extensively.

made in the world oil market that share the common elements of these alternative policy options. Growing demand, increasing oil-related investments, balance of trade deficits for consumer countries, and periods of capital inflows will persist. But there will not always be surplus oil revenues, nor always the resources for domestic investments for development. As at the present time corporate profits will be tied to the volume and price of oil trade, and the trade will continue to determine required investments in the oil industry.

At this writing "Breaking the cartel," a popular cliche during the early 1970s, is no longer heard as much in academic or policy-making circles. Threatening embargoes and price wars are no longer at the forefront of the producer countries' strategies, or even rhetoric. The underlying analysis of this book can be viewed as a set of forecasts, using scenarios as a means of reducing uncertainty about the future. We cannot see the future, but we can show the consequences of various policy options by the producer countries.

There are always trade-offs for each country as it responds to the oil market. The costs and benefits created by alternative tax policies reveal the broad effects for the participants in the exchange.

All tax policies will entail some adjustment in demand, but only a radical policy will impose strong downward pressures. All tax policies will result in growth of consumer oil payments, but only the radical and conservative policies can allow for some years of actual leveling of import expenditures. As more investments are directed toward energy resources of the consumer countries, the future can be considered more favorable for them than the past or present. In the long run "Breaking the cartel" or returning to 1970 prices will reduce the incentives to invest in substitute energies and push consumers toward ever greater dependence on diminishing oil reserves.

For the producer countries the future represents the clearest options. A radical tax policy will produce the greatest revenues. But the trend will taper off, as consumers attempt import cuts. If the oil producers want the largest revenues, subject to the constraint of consumer responses, they will opt for the radical policy. If they want a steady growth in revenue, but at lower levels, the optimal policy is most desirable. A breakup of OPEC by 1985 will cut revenues and will produce a situation that will yield lower revenues over time than the other policies. The radical case will also create a sharp increase in the producers' foreign investments, followed by a decline due to consumer cuts in imports. The optimal case will create a steady stream of foreign investments, but at lower levels. This is one important key to future interdependence.

Since corporate profits are tied to the volume of trade, there are three cases in which the oil companies will be worse off than they are under present circumstances: a radical tax policy, a conservative policy, or an optimal policy. In all other cases the future portends growth in corporate profits.

These are all contingencies. But underlying these contingencies are some robust forecasts, strident statements about the future. A gradual increase in the tax rate will allow for systemwide adjustments. Any sharp shock will take longer for the effects to reverberate throughout the network of interactions. Of the policies we have examined, the conservative and radical options—or variants of them—provide the bounds for our best guess of things to come. The gains to the producers are greatest, and the possibilities of consumer adjustments also extensive. In any event, whatever the price or the volume of trade, the consumers will face heavier burdens in oil payments. Higher oil prices will create greater costs for consumers, but in the long run they could be easier to bear than the overall political and economic costs triggered by "Breaking the cartel." Low oil prices have been responsible for the absence of a coherent energy policy in the West and for the lack of foresight by the producers. Higher prices will impose constraints on everyone and force all parties to the exchange toward more rationalized policy responses.

Choices can, must, and will inevitably be made. Debates on energy policy are efforts to articulate preferences and the willingness to incur costs or reap benefits. The policy problem for each participant in the world market is to determine the extent of costs, the hierarchy of preferences, and the policies available to assist equitable exchanges of global energy resources at acceptable costs.

Policies come from action. Action must be based on vision. Short-term policies are invariably myopic. Costs and benefits viewed exclusively in narrow economic terms are misleading. National policies are based on political and economic factors. Power and money may not be interchangeable, but they are interconnected. The world is not only a market place—it is also one of conflict, bargaining, and political exchange. The future that will unfold tomorrow, the day after, and years beyond will draw upon the past and will have common elements.

2
The International Petroleum Exchange Model

A model is an abstraction of reality that depends on purposes, assumptions, and analytical tools, and skill and ingenuity in approaching the issues. Generally economists concur that supply and demand relationships constitute the defining characteristics of most international markets, but there is less agreement as to the best means of characterizing such relationships in the case of the oil market. The analytical structure of noncompetitive markets can be used to examine certain aspects of the oil market, but unlike the competitive market model its properties are not fully tractable. With few exceptions the international oil market where power politics strongly influences the supply of a commodity remains beyond the pale of formal economic theory.[1]

Most oil models seek to compare OPEC-generated prices with those likely to prevail under competitive conditions. Implicitly, if not explicitly, the OPEC phenomenon is regarded as an aberration from normal conditions, with the solution to market imperfections being a return to competitive conditions. Competition, however, is only a useful idealization and seldom does it even approximate reality. While many studies succumb to the temptation to treat oil exchanges as if they should be competitive ones, the conditions in the world oil exchanges prior to October 1973 were not characteristic of a competitive market.

2.1 Dynamic Analysis

The International Petroleum Exchange (IPE) model focuses on price determination and its worldwide effects. The parties in the exchange are aggregate buyers (the oil-consuming and importing countries of the OECD), aggregate sellers (the oil-exporting countries of the Gulf region in the Middle East), and international managers (the major oil companies).[2] Sellers besides those in the Gulf are not treated explicitly but are assumed to play a role in market equilibration in the short run.[3] In 1970 the Gulf region accounted for 28 percent of all petroleum production and 51 percent of world exports. In 1978 the figures were 33 and 58 percent. At that time the OECD countries represented 82 percent of all worldwide imports and 79 percent of imports from the Gulf.

The IPE model gives a detailed description of the market, taking into account the interdependence generated by multiple sources of demand for oil, by constraints on supply, by the nature of the operating rules and regulations that relate supply to demand, and by the attendant price adjustments. It is structured as a dynamic simulation initialized at 1970 values and recalculated incrementally to 2000.[4] Setting initial values (and key parameters) to 1970 is designed to delineate the structure of exchanges prior to the price increases of October 1973. The intent is to trace the effects of this disturbance in the market and compare the resulting changes since 1970 to the changes that might occur with alternative oil prices.

The major features of the IPE model are the following:

1. The model adopts a *political economy* perspective that includes, but extends beyond, the confines of one market and takes into account oil production processes, oil trade, and international financial and security consequences.
2. It is structured in terms of interactions among *three relevant entities*—producer countries, consumer nations, and international oil companies.
3. A component of *price* is set by the exporting countries in their determination of the tax rate on extraction, but the importing nations and the international oil companies also influence price.
4. Price is a function of the tax rate, oil production *costs,* and the markup of the international oil companies.
5. *Markup* is a means by which the oil companies adjust to supply and demand influences in the world oil market.
6. The quantity of oil supplied is determined largely in terms of oil *production* in the exporting countries; however, there is provision for the use of domestic sources of oil in the consuming countries.
7. *Demand* is formulated in terms of total consumer demand for oil and demand for imports from the Gulf area.
8. *Imports* from the Gulf are calculated taking into account domestic sources of petroleum in the oil-importing countries.
9. Imports and domestic production are influenced by the price of oil that also determines the extent to which *energy substitutes* become available.
10. Imports from the Gulf generate *oil payments* that contribute to the

producer countries' revenues and appear as a major claim against consumer countries' balance of payments.

11. The *balance of payments* is computed for all petroleum-related transactions—oil payments to the exporting countries, the investments of the oil producers in the economies of the consumer nations and their purchases of goods and services from the consumers, as well as the repatriation of profits by the international oil companies.

12. Fundamentally the model is one of *international interdependence* reflected in the economic, resources, and political interactions underlying oil trade.

The emphasis is on the price of crude petroleum; however, the price of refined products to the consumer countries can be specified with minor changes in the model. But the price determination process fundamentally begins with transactions in crude oil; the price to the consumers of oil-related products must by necessity be predicated on the price of crude.

2.2 Analytic Representation

The element of time is essential in a dynamic analysis of the world oil market. The oil-exporting countries influence prices by setting the tax rate, but at the same time their influence is constrained by the supply and demand relationships that impinge upon price. Price is set as a function of the tax rate, the costs of production, and the markup of the oil companies. Price changes affect both the quantity demanded and the amount supplied. In turn supply and demand adjust to price. However, there are time lags involved on both sides. On the supply side there are the lags associated with investment delays. In the short run demand adjusts to price, and supply from the Gulf is relatively unresponsive. Non-Gulf supplies adjust to meet demand at the prevailing price. Over the longer run both supply and demand adjust to price and in turn influence the final determination of price.

The activities and interactions of the producer countries, the consumers, and the international oil companies generate the adjustments of supply and demand to price. These adjustments influence price by altering the markup of the oil companies. The tax rate of the producer countries is exogenous. But tax is only one component of price. Depending on the size of the tax rate, it is not always the largest component.

These adjustments to price are not instantaneous. Price at t_1 leads to a quantity demanded, which leads to an amount supplied. That amount is constrained by previous demand patterns and by the costs and investments that have generated productive capacity. Demand in turn is influenced by past prices and by price expectations. Depending on price and quantity, the process may be extremely stable. But with rapid changes in the tax rate there is a dynamic interaction that may induce instabilities in the process because of the lagged response of both supply and demand.

Figure 2.1 highlights these relationships in the short run. Gulf supply is unresponsive to price, non-Gulf supplies adjust to demand, and demand adjustments due to price changes are made. The vertical supply curves indicate the short-run fixity of Gulf supply. Over time both supply and demand adjust to changes in price. Figure 2.2 indicates the dynamic interaction. Note that the quantity supplied responds both to demand and to price. Conversely the quantity demanded responds to price and supply. The driving mechanism in the model is for supply to adjust to the amount demanded which in turn responds to price. In figure 2.2 the driving mechanism is a supply curve shift corresponding to the producers raising the tax rate. In effect an increase in the tax means the producers will sell

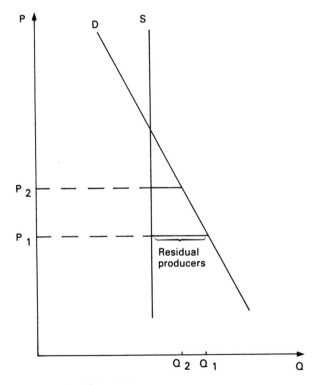

Figure 2.1 Short-run adjustment process

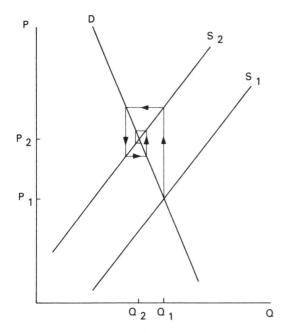

Figure 2.2 Longer-run adjustment process

the same quantity only at a higher price, hence the supply curve shifts to the left. The result is a new equilibrium at a higher price and a lower quantity than before, with residual, non-Gulf suppliers acting to clear the market during the transition. This figure of course does not represent the dynamic, time dependent process, nor does it show the prices and quantities supplied and demanded at a given time period. It does indicate, however, the type of supply and demand interactions modeled for the oil market and the pressures that may exist under certain conditions for a new equilibrium.

We do not seek to specify the price-determining process in its full complexity by modeling the decision process by which suppliers set their tax rate. However, we seek to specify and include major factors that generate a final price of crude oil and then to examine international economic and political consequences of alternative oil prices. The impacts of price upon key variables in oil trade, such as consumer demand, expenditures on oil imports, balance of payments, corporate oil profits, and producer revenue are modeled explicitly, as are the alternative financial investment opportunities of the producer countries. This model of price determination is one in which there is an interdependence of supply and demand, but that interdependence is not the only influence on price.

The model is based on key equations that represent dynamic behavior and a set of accounting equations that monitor the effects of this behavior. The essential features, of course, are the demand, supply, and price relationships. The model is written in a simulation language and is not readily formulated in econometric terms. Conceptually, however, the essential features of the model can be summarized and are given in table 2.1, with the caveat that this representation is for analytical purposes rather than to depict the precise relationships. The table presents the core of the IPE model. The left-hand variables are endogenous as are some of the right-hand variables. Once quantities supplied and demanded are generated—at a particular price—consumer balance of payment, investments and imports, and corporate profits and investments in the oil industry are computed. Over time then the dynamic adjustments in the model are of supply and demand to price changes. Final price of course is responsive to supply and demand conditions.

Demand
Quantity demanded is modeled as a dynamic process, adjusting to prices, to substitutes, and indirectly to income. The demand schedule for the commodity reflects the demand of all consumers, that is, those that import petroleum as well as the consumption of the sellers. However, the functional form for each group is different in the degree of specification and detail. Total demand is then a sum of the demand of buyers and of sellers.

In the case of the buyers, the oil-consuming countries of the West, demand is a function of forecasted OECD demand, adjusted by substitute availability, domestic oil production, and price. The OECD calculations are themselves based on income assumptions and expected price.

The effect of energy substitutes upon demand is represented as follows: we begin with a projection of future prices of substitutes and compare oil price to the projected substitute price at that point in time. Substitute availability is determined by the price of substitute energies in any given year. The higher the price of oil, the greater the percentage of oil demand covered by substitutes. The final output of these calculations is the influence on demand. Imports demand is then computed residually, as the difference between demand and production from domestic sources.

For the demand of the sellers, the oil-exporting countries, regression estimates have been obtained based on the impact of population and capital stock on oil consumption. The estimates are derived from pooling cross-sectional and longitudinal data for key Gulf countries.

Table 2.1
Analytical representation of the IPE model

	Key equations	Variables
Supply	$TS = S_p + S_c + S_R$	$TS =$ total supply
	$S_p = f(PC, R)$	$S_p =$ producer supply from Gulf
	$S_c = f(\hat{S}, P, DR)$	
	$S_R = TD - S_p - S_c$	$S_c =$ consumer supply
		$PC =$ production capacity
		$R =$ reserves
		$\hat{S} =$ base OECD series
		$P =$ price
		$DR =$ decline rate
		$S_R =$ output of residual sellers, inventory changes, etc.
Demand	$TD = D_c + D_p$	$TD =$ total demand
	$D_c = f(\hat{D}, P, E)$	$D_c =$ consumer demand
	$D_p = f(Pop, K)$	$D_p =$ producer demand
		$\hat{D} =$ base OECD series
		$E =$ alternative energy sources
		$Pop =$ producer population
		$K =$ producer capital stock
Price	$P = T + C + MK$	$T =$ tax rate
		$C =$ cost
		$MK =$ corporate markup
Markup	$MK = f(MK_0, PCU, DR, \triangle COMD)$	$MK_0 =$ base markup
		$PCU =$ production capacity utilization
		$\triangle COMD =$ change in consumer import demand
Consumer import demand	$COMD = D_c - S_c$	
Market clearing	$S_R = COMD + D_p - S_p$	

Supply

Supply is modeled explicitly for an entire production process, taking into account physical conditions, technological factors of exploration and development, and economic ones of costs and price. Specification of domestic supply in the consumer countries is written less comprehensively but takes into account price and depletion.

The supply of oil from the exporting countries is made in terms of the relationship between reserves and oil production. Production is affected most directly by demand, given adequate levels of reserves and production capacity. While reserves of oil in the ground are in the producer countries, as modeled, investments in exploration and development are undertaken by the oil companies. Exploitation of these reserves is the basis of the supply of petroleum imported by the consumer countries. Effective exploration investments determine the oil discovery rate that transforms undiscovered oil (initialized at 500 billion barrels for the Gulf fields) into recoverable oil.[5] Investments in development and production costs determine production capacity. The ratio of production to production capacity sets the amount of capacity utilized.

In this model there is no explicit specification of inventory, or inventory responsiveness to changes in price, in quantity supplied or demanded. Changes in the stock of recoverable oil-in-place occurs with variations in the oil discovery rate and in production. Recoverable oil-in-place in producer countries varies inversely and nonlinearly with output.

Modeling consumer production is designed to explain the main alternative source of supply other than the Gulf. Consumers may draw upon domestic sources of oil by increasing their own production. But, since known reserves in consumer countries other than the United States and those in the North Sea region are not extensive (and are economically unattractive given present technology), it is not unrealistic to stipulate a physical constraint on consumer oil production. These reserves are initialized at 200 billion barrels and can result in depletion within the time horizon of the model. Recoverable reserves in the consumer countries provide some constraints on price setting.

Supply from domestic sources in consumer countries draws on base OECD projections adjusted by price and the rate of depletion. Supply elasticity, indicating the responsiveness of quantities offered for sale to a change in price, as a long-run coefficient set at 0.3. In the short term of course supply is relatively inelastic.[6]

This formulation of supply represents an important aspect of the interdependence among three major entities in international petroleum exchanges. Consumer oil import demand and producer demand for oil result in total demand for Gulf oil. Managers of energy exchanges, the multinational oil corporations, influence indirectly the production process through their investments in exploration and development and the margins they set which in turn influence industry parameters and the size of the reserves.

Price

Price is a function of the producers' tax rate, the costs of production, and the corporations' markup. For the producer countries to increase the tax rate obviously results in increases in price and an immediate increase in their own revenues, given inelastic demand. But it enhances the incentives for consumer countries to reduce their demand for Gulf oil, partly by increasing investments in substitute energies. This consumer reaction constitutes one of the competitive elements of the oil market that is modeled explicitly through the determination of corporate markup.

Corporate markup takes into account capacity utilization in the production process, the impact of the decline rate, and the gap between the consumer countries' demand for imports from the Gulf and the Gulf's ability to meet that demand. Markup represents the dynamic, endogenously determined influences on price. Through markup the systemwide supply and demand conditions influence price, and in turn prices influence the prevailing conditions in the oil market as a whole.

The dominant role of the producers in setting prices should not obscure the interaction of supply and demand. When the quantity demanded changes, an adjustment in supply occurs, and both changes influence price. Analytically a simultaneity is implied, and in terms of the computations the adjustment process is an interactive one. This view of price determination does not suggest that the quantity purchased is entirely independent of influences other than price or that there is considerable price inelasticity. Rather it acknowledges the special role of the producers in setting the tax rate on the barrel of crude oil and renders explicit the supply, demand, and cost constraints within which they operate.

Under alternative assumptions, or "scenarios," other components of price could become more influential. If prices were presently at the pre-1973 level, cost would be a higher proportion of price than it is today. Of course the cost of production may increase in the years to come in most Gulf fields. Markup, which represents the influence of international oil companies, reflects their attempt to adjust to the prevailing supply and demand conditions in the oil market. Depending on market conditions and the tax rate, markup could be a large component of price. The consumer countries have a direct impact upon the oil market by exercising some leverage on demand, on the domestic sources of petroleum and the prospects of substitute availability. Of course in a price-setting process where the price of end-use products is at issue, the taxes of the consumer governments become a critical component of final price.

The effects of price reverberate throughout the entire model, influencing the exporting countries and the importing countries separately. The exporters are affected directly by oil revenue, production, capacity utilization, and so forth. The importers are affected by demand, domestic oil production, quantity imported, and the availability of substitutes. In these respects there is a dynamic interaction between supply, demand, and price.

Adjustment Mechanisms
The agents in the market can engage in activities that contribute to the adjustment process in both the short and the longer run. The IPE model combines the characteristics of two types of economic models—the dominant firm model for a short-run analysis and the longer-run adjustments of quantity to price.[7] The dominant firm model applies to the Gulf producers who make price and quantity decisions in the short run by setting the tax rate on extraction and/or the amounts to be produced or capacity to be utilized. In the longer run their production responds to the size of the residual market (where excess demand is met in the short run by non-Gulf producers) and to the price responsiveness of demand. Less directly relevant to the oil market, but important for the overall oil-related transactions, are the investment policies of the producer countries.

In the short run consumers can influence the size of the residual supply, increase domestic production, and to some extent cut their imports. In the longer run they can reduce demand and expand the use of alternative sources of energy.

The companies' markup is the immediate adjustment to prevailing market conditions. In the longer run they influence exploration and development through investments in the oil industry. Their impact on Gulf supply is thus of a longer-run nature.

Due to the complicated nature of the model it would be very tedious to evaluate its stability properties in analytical detail. The longer-run adjustment is a cobweb model with apparent price convergence.

2.3 Computation Structure

The IPE model is composed of seven sectors. It is designed to represent the physical characteristics of oil production, the economic context and constraints, and the international financial exchanges that ensue from the trade in oil. Figure 2.3 represents conceptual relationships, and not the computer readable equations. It summarizes the model as a whole, in that producers and consumers interact through financial and economic transactions, mediated by the activities of the international oil companies and constrained by the geological and technological features of the oil production process. The theory of price formulation of this model is given in summary form in figure 2.3.

The sectors of the model can be described briefly as follows:

• The *supply sector* represents the physical stages of oil production, tracing the process from exploration for oil-in-place and the development of recoverable reserves to the installation of productive capacity and actual production.

• The *finance sector* makes key calculations for each of the three entities in the oil market: oil import expenditures for consumer countries, corporate profits and oil investments for the oil companies, and oil income for the producer countries.

• The *management sector* specifies the corporate investment decisions affecting the supply of oil. The major investments of the multinational corporations in development and exploration are based on information drawn primarily from the supply sector, in conjunction with considerations of oil demand from the consumer sector.

• The *price sector* calculates the price of oil based on inputs from other sectors of the model. The tax rate is the exogenous component; production costs and the corporate markup are also taken into account. Once calculated, the effects of price are then transmitted throughout the model to compute its financial and security implications for producers and consumer.

• The *producer sector* models the process of industrial development in the Gulf states, which generates demand for development investment and for imports of goods and services. The tax rate is set in this sector as an external policy variable. It is a key input of the price calculations.

• The *consumer sector* computes demand for oil imports and monitors the consequences of such imports for the consumers' strategic vulnerability and dependence upon external sources of supply. This sector models supply and demand for oil from domestic sources in consumer countries and the availability of substitutes.

• The *international economic sector* calculates the consumer balance of

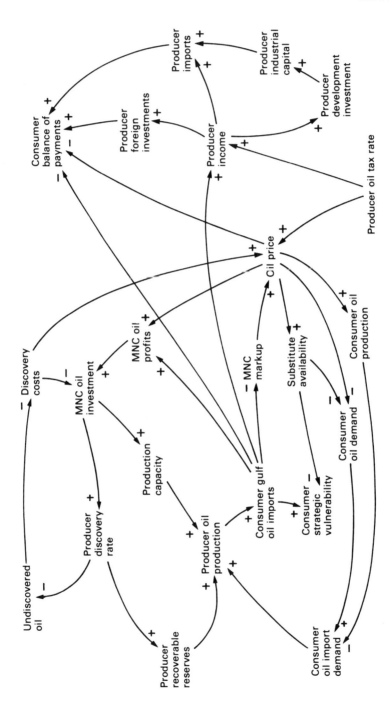

Figure 2.3 Simplified model overview of major causal loops

payments and traces the foreign investment activities of the producing states; it registers and links the consequences of actions taken by the consumer countries, the producers, and the oil companies.

The Supply Sector

The supply sector represents the oil production process, from initial exploration investments to the development of oil pools and the generation of output. This specification follows the description provided by Adelman (1972).

Oil production is modeled explicitly in terms of exploration and discoveries, capacity utilization, decline rate of existing fields, and discovery and production costs. Each of these factors reflects a certain extent of interaction among consumer countries, producer countries, and international oil companies. Although known reserves in the Gulf region are estimated to be ample at least until the turn of the century, given the time horizon of the model, it is reasonable to inquire into the long-term implications of different production rates, levels of oil investments, or the decline rates of oil production in the region.

The two key sets of relationships in this sector generate oil reserves and production:

$$\text{Producer reserves}_t\ (\text{PROIP}) = \text{Reserves}_{t-1}\ (\text{PROIP}) - \text{Production}\ (\text{POP}) + (\text{Discoveries}\ (\text{PODR}) \cdot \text{Recovery fraction}\ (\text{ORFI})),$$

$$\text{Producer production}\ (\text{POP}) = \text{Min (Total demand for Gulf oil}\ (\text{TGOD}),\ \text{Production capacity}\ (\text{POPC}),\ \text{Reserves}\ (\text{PROIP})).[8]$$

The supply sector generates oil production capacity and estimates how much is actually utilized. It also specifies the reserves and remaining undiscovered oil. The oil companies' effective investments in the oil fields are computed, as well as the amount of oil exported to consumer countries.

The Finance Sector

The three critical features of world oil trade are specified in this sector: (1) corporate profits and investments in the international oil industry, (2) the expenditures of the consumer countries in payment for petroleum imports from the Gulf, and (3) the revenue of the exporting countries from the oil sales.

It is extremely difficult to model corporate profits accurately or determine their investments in the oil industry. In this model corporate oil

profits are a function of the consumer countries' oil import expenditures,
the costs of oil production (reflecting the cost constraints on profit), and
the oil producers' oil taxes and royalties:

Profits = Consumer oil   − / Oil production + Taxes and \ .
(MOP)   import expenditures (cost   royalties)
           (COME)   \ (MOC)   (POT)   /

Oil production cost is computed as the per barrel production cost
multiplied by total barrels imported by consumer countries. The total cost
per barrel is specified as the sum of exploration costs and development
costs, which include operating and recovery costs.

Consumer oil import expenditures are simply the quantity of oil imported times the price per barrel:

Consumer oil import expenditures = Price · Consumer Gulf oil imports.
(COME)   (P)   (CGOM)

Oil income, or revenue, is determined by taxes and royalties less the
total cost of producing those barrels of oil consumed by the producer
countries themselves:

Producer oil income = Taxes and − / Cost per · Producer oil \ .
(POY)   royalties   (barrel   consumption   )
                      (POT)   \ (PMOC)   (PC)   /

The distinction between the profits of the oil companies and the income
of the oil-producing states accrued from the sale of oil reaffirms the dis-
parate roles and gains from oil sales under different price and market
conditions. Corporate profits in this model provide a source of finance for
investments in the oil industry. The total investments in that industry are
derived from two sources: the reinvested profits of the oil companies and
the investment allocations of the producer countries.

The Management Sector
The investment decisions made by the international oil companies as they
evaluate demand conditions (in terms of desired additional capacity and
what is to be produced) and supply conditions (in terms of desired discoveries and depletion targets) are specified in the management sector. These
decisions determine exploration and development investments and corporate markup on price.

Exploration investments influence oil reserves through the discovery
rate. Development investments influence production capacity. In each
case there are time lags between the actual investments and their effec-

tiveness. The decisions governing investment requirements take account of desired discoveries, discovery and production costs per unit capacity, desired additional production capacity, and oil capital depreciation. Desired additional production capacity is determined as follows:

Desired additional	= Desired production	− Production	+ Production capacity
production capacity	capacity	capacity	loss to depreciation.
(DPOPCA)	(DPOPC)	(POPC)	(POPCD)

As modeled, the oil exploration investment begins with the oil companies' comparison of actual and desired decline rates. The decline rate is the rate at which reserves are depleted, given prevailing levels of oil production. If proven reserves are being depleted more rapidly than desired, then the companies will seek to discover more oil. Conversely, if reserves are being depleted at a lower than desired rate, then the companies will try to adjust their oil discoveries so that the decline rate rises accordingly. The amount of discoveries desired are computed in the following manner:

$$
\begin{array}{ll}
\text{Desired} & = \text{Production} \left(\text{Decline rate} \div \text{Desired decline rate} \right) \\
\text{discoveries} & \quad\ (\text{POP}) \quad\quad (\text{POPDR}) \quad\quad\quad (\text{DPOPDR}) \\
(\text{DDIS}) &
\end{array}
$$

The producer countries have direct inputs into these decisions. They may set (exogenously) the desired decline rate and desired capacity utilization, since these are essentially policy variables. Both variables are important inputs into exploration and development investment decisions. Despite major changes in the world oil market since 1970, the base year of the model, the oil companies continue to play an important role in the management of the industry internationally.

The Price Sector
The price computations are based on information generated in other sectors. Price is determined by the producers' tax rate, the cost of production, and the multinational companies' markup. The final price per barrel is subject to constraints modeled endogenously, including supply and demand relationships. The noncompetitive elements of the oil market are represented by the tax rate of the producer countries, and the competitive elements are represented by the supply and demand relationships embedded in markup.

The price equation is as follows:

Price	= Tax rate	+ Costs	+ Markup.
(P)	(POTR)	(PMOC)	(MNCM)

The tax rate can be thought of as the rent paid to owners of productive resources. The underlying goals and objectives of producer countries are implicit in the tax rate. These goals may include maximizing present value of some long-run income stream, meeting certain political objectives, or maintaining particular production levels. We do not model these goals explicitly but recognize the complexity of the decisions governing the tax rate.

The cost component of price entails discovery and recovery costs. Both are specified explicitly. Markup is influenced by market conditions and the multinational oil companies' investment decisions. Markup is a device by which the managers of petroleum exchanges—the oil companies— will attempt to adjust the relationships between supply and demand, by marginally modifying the price. Markup is specified as follows:

$$\text{Markup} = \begin{pmatrix} \text{Base} & + \text{Capacity} \\ \text{markup} & \text{utilization} \\ \text{(CMI)} & \text{discrepancy} \\ & \text{(POCUD)} \end{pmatrix} \cdot \begin{pmatrix} \text{Decline rate} \\ \text{effect} \\ \text{(POPDRM)} \end{pmatrix} \cdot \begin{pmatrix} \text{Consumer import} \\ \text{gap effect} \\ \text{(COMGM)} \end{pmatrix}.$$
(MNCM)

The base markup is set at $.82, which is the 1970 markup on Saudi light crude. The other three components are functions of discrepancies between demand and either actual or desired supply. All three variables reflect demand and supply influences.[9] Desired capacity utilization is set at 90 percent of capacity, and the desired decline rate is specified at 10 percent per year, and the gap indicates market tightness. These calculations are designed to approximate the adjustment mechanisms of the world petroleum market, representing corporate decisions to restore equilibrium between demand and production by increasing or decreasing markup.

The Producer Sector
The oil-exporting countries are modeled in terms of three sets of relationships: investment demand for their own development, demands for imported goods and services, and demand for domestic oil consumption. Each relationship is partially dependent on the revenue obtained from the sale of oil, population, and the existing industrial capital stock.

Population level is initialized at the 1970 level for the Gulf states, and specified to increase by 3.1 percent per year, a weighted average of the 1970 population growth rates of these states.

The producer countries' capital stock is set as a function of the past rate of capital investment and of the rate of capital depreciation. Setting the initial level of capital stock is not easy, given the uncertainties in available data for developing countries. A rough approximation commonly employed by development economists is used in setting the capital stock at 2.5 times the gross domestic product for Gulf states.

The three key demand equations of the producer countries are as follows:

Producer capital $= \alpha_1 + \beta_1$ Capital stock $+ \beta_2$ Oil income
investment demand   (PK)   (POY)
(PKID)

$+ \beta_3$ Population $+ \mu_1$,
        (PPOP)

Producer import $= \alpha_2 + \beta_4$ Population $+ \beta_5$ Capital stock
demand   (PPOP)   (PK)
(PTMD)

$+ \beta_6$ Oil income $+ \mu_2$,
        (POY)

Producer oil $= \alpha_3 + \beta_7$ Population $+ \beta_8$ Producer capital stock $+ \mu_3$.
demand   (PPOP)   (PK)
(PCD)

The producer entity represents a generic supplier with the aggregate combined characteristics of Gulf states. The coefficients are estimates by pooling cross-sectional and time series data for Iran, Saudi Arabia, and Kuwait.

The Consumer Sector

The consumer sector generates consumer oil imports and monitors the strategic implications of these imports. The specification of the consumer entity as a generic consumer may invariably obscure the differences among individual consuming nations and the extent to which each is vulnerable to interruptions in oil supplies, but broad patterns can be delineated.

The demand for imports occurs when domestic consumption exceeds domestic production. To calculate imports, it is necessary to formulate the domestic production relationship (that component of supply that refers to domestic sources of oil and is controlled by consumer countries) and domestic oil demand (indicating total consumer demand for oil).

Consumer oil production takes into account the known consumer oil reserves, the impact of oil price, and depletion effects. Oil production is based on OECD projections, adjusted endogenously by expected price and depletion as follows:[10]

Consumer oil   = Exogenous forecast of · Effect of price · Effect of depletion
production   consumer production   on production   on production.
(COP)   (COPT)   (CPMOP)   (CDPL)

The OECD forecasts of oil production are estimates growing from 4.2 billion barrels per year in 1970 to 6.1 billion barrels in 1985, based on a forecast of $3.00 per barrel.[11] This price is extremely low by current standards. But, since the model is initialized at 1970 values, the OECD estimate is a good point of departure. The adjusted production forecast enables consumers to draw upon domestic sources of petroleum but recognizes that these sources are constrained and that reserves are not extensive or as economically competitive as Gulf sources. These constraints provide the rationale for calculating oil demand and import demand.

Consumer demand for oil is a function of the interactive effect of forecasting demand (taken from the same OECD projections, again assuming a price of $3.00 per barrel), the effect of current price upon demand, and the impact of substitute energies upon demand. The gap between demand and production from domestic sources determines the demand for oil imports. Since the effect of substitute availability is incorporated in the calculation of consumer demand, when the difference between consumer demand and domestic production is computed to determine import demand, the impact of alternative energies has already been taken into account.[12] Thus

Consumer   = Exogenous demand · Price effect · Substitute availability
oil demand   forecast   on demand   effect
(COD)   (CODS)   (CPMOD)   (COSAM)

and

Consumer oil   = Consumer oil demand − Consumer oil production.
import demand   (COD)   (COP)
(COMD)

The effects of price on demand are both direct and indirect. Price affects the quantity of petroleum demanded directly by a function that relates the projection of consumer oil demand, assuming a constant price of $3.00 per barrel, compares the current expected price (an exponential average of past prices from the previous three time periods) with the IPE model base price of $1.80 per barrel in 1970, and adjusts demand accordingly.

Price affects the quantity of petroleum demand indirectly through its effects on consumer oil production. As the price per barrel increases, consumer countries increase their domestic production of petroleum; as their domestic production increases, their demand for Gulf petroleum declines (assuming all other factors remain constant).

Price affects quantity demanded indirectly also through petroleum substitutes. As price per barrel increases, consumer oil demand is increasingly met by substitute energy. The price of one-barrel equivalent of petroleum substitute is specified to decrease over time, reflecting the assumption that technological development will gradually result in the reduction of the cost of substitute energies.

The consumer countries' vulnerability associated with oil trade is modeled in this sector. The vulnerability is viewed in strategic terms, reflecting four distinct, but interrelated, factors: (1) dependence upon external sources of oil supply, (2) criticality of oil to industrial economies, (3) availability of substitutes and (4) insecurity of oil supplies.

Dependence on external supply is the ratio of imports to total consumption. Essentiality of oil is the ratio of oil consumption to total energy consumption. Substitute availability is related to the development of alternative energy technology and the price of oil. Insecurity of supplies is the ratio of imports from the Gulf relative to total oil demand of the consumer countries. These indicators yield a combined vulnerability index that states that consumer countries are more secure as they import less oil from Gulf sources and rely less on oil overall. The composite strategic vulnerability index thus monitors the overall effects of international transactions in petroleum for consumer countries.

The International Economic Sector
The balance of payments calculated in the international economic sector shows the influences of oil trade on international monetary transactions. The current account represents all trade in goods and services, including government exports and grants. The capital account includes private and government short- and long-term loans and equity investment transactions. The basic balance conception of international balance of payments equilibrium stipulates that total exports minus total imports plus long-term capital flows be equal to zero. The important issues with respect to balance of payments equilibrium are whether existing flows can be maintained and whether trade exchanges indicate basically sound transactions or marked imbalances.

The computations are done from the perspective of the consumer countries; the effects are systemwide. Since the focus is on oil trade, the balance of payments is modeled between the consumers and producers, along with some factors that influence the balance of payments or the instruments available for importers or exporters to redress imbalances.

The current account is set as the consumers' trade balance plus the difference between repatriated corporate profits and repatriated income from producer countries' investments overseas:

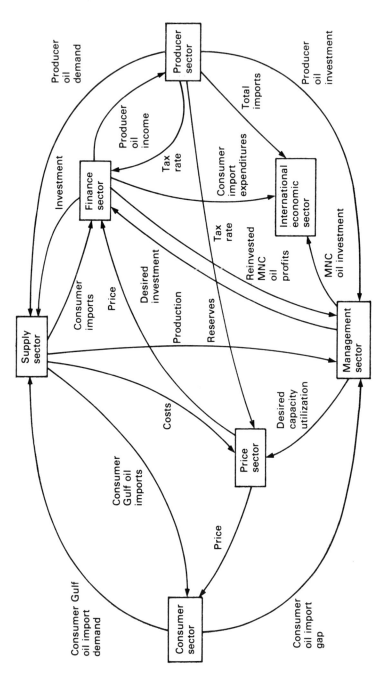

Figure 2.4 Major linkages of the sector link diagram. Note that only one component of markup is derived in the management sector (see chapters 5 and 6).

$$\begin{aligned}
\text{Current account} &= \text{Consumer oil} + \left(\begin{array}{l} \text{Corporate profits} - \text{Producer repatriated} \\ \text{repatriated} \qquad\quad \text{investment income} \\ \text{(MRPOP)} \qquad\qquad \text{(PRIY)} \end{array} \right) \\
\text{(CBCA)} &\qquad\quad \text{trade balance} \\
&\qquad\quad \text{(CBT)}
\end{aligned}$$

The consumers' oil trade balance is the difference between the consumers' oil import expenditures and the producer countries' total imports of goods and services from the consumer countries.

The capital account is set as the difference between the producer countries' investments in consumer countries and their investments remitted to their home countries, less the multinational corporations' new commitments to oil investments in producer countries:

$$\begin{aligned}
\text{Capital account} &= \left(\begin{array}{l} \text{Producer countries'} - \text{Producer countries'} \\ \text{foreign investments} \quad \text{investments remitted} \\ \text{(PFI)} \qquad\qquad\quad \text{(PIR)} \end{array} \right) \\
\text{(CBKA)} &\qquad\quad - \text{ Oil companies'} \\
&\qquad\qquad \text{direct investments.} \\
&\qquad\qquad \text{(MDOI)}
\end{aligned}$$

The consumers' basic balance equals the capital account plus the current account, while a fourth balance of payments variable, labeled the cumulative basic balance, is devised to monitor the consumers' overall payments over time. It is set to zero during the first simulation period, and then accumulates, revealing the long-term effects of petroleum trade.

From the producer countries' point of view the predicaments posed by oil revenue have been referred to as potentially (or partially) resolved by recycling petroleum revenues. The producers' foreign investments are a direct input into the consumers' capital account, and the producers' imports of goods and services are an input into the consumers' balance of trade. These capital specifications highlight the interactions among the major entities in petroleum exchanges and reflect the financial effects of oil trade.

Changes in structural characteristics of each entity, and in differential power and capacity, can be inferred from observing the evolutionary behavior of output variables under alternative scenarios and underlying assumptions. The key issue is who gains under the different price and production policies. This issue sheds light on the oil market and the conditions of international petroleum exchanges at any point in time.

Figure 2.4 depicts the general computational structure of the IPE model. It gives the connections among the sectors and the key variables providing the links. It is a guide to the design of the simulation model and its overall integration.[13]

II
The Model

3
The Supply Sector: The Process of Oil Production

In formulating the equations that represent oil production, we specify the phases of the process, from the initial exploration investments to development of oil pools and actual production. The modeling process reveals the demand and supply mechanisms that influence interactions among producers, consumers, and oil companies.

Exploration and development investments are required to find deposits to generate reserves and to create production capacity. Bringing new pools on-line in essence involves four basic variables:[1] (1) investments, such as in exploration and development, (2) production capacity, indicating the extent to which oil can be extracted, (3) decline rate, measuring the extent of depletion of oil reserves within a country or region, and (4) production costs, involving expenditures on day-to-day operations. To varying extents these four variables reflect the interactions among the three major entities in petroleum exchanges. Producers, consumers, and companies have a different degree of control over the decisions that set these parameters, but each exerts some direct or indirect influence over these four parameters.

3.1 Description of the Production Process

Modeling the process of oil production—with the Gulf region as the empirical referent—entails writing a set of equations representing the generation of oil output. The equations are based on Adelman (1972), who describes the production process and gives supporting evidence necessary for formal representation of the decisions that govern petroleum production.[2]

Oil production can be characterized as a three-stage process: (1) investing in exploration to locate oil-in-place, (2) investing in development to create productive capacity, and (3) lifting oil from the ground.

Exploration and Discovery
In the Gulf region it is estimated that the magnitude of recoverable oil-in-place in 1970 was 350 billion barrels, and the reservoir of undiscovered oil, both recoverable and unrecoverable, has been estimated at 500 billion barrels. The purpose of exploration and discovery is to increase the size of recoverable oil-in-place. This process depends on the discovery rate and on the oil recovery fraction. The discovery rate in turn is dependent upon the amount invested in exploration and the cost of discovery. Discovery costs depend upon the magnitude of undiscovered oil or the underlying deposits. The decline rate in recoverable oil-in-place is critical in determining how much will be invested in exploration. It is conventionally viewed as a function of production, in that, the higher the rate of production in a particular field, the greater will be the decline rate, thus reducing the size of known recoverable oil-in-place.

The exploration stage can be formulated as a negative feedback loop reflecting the transformation of underlying reservoirs into recoverable oil-in-place (usually termed proved reserves).[3] Figure 3.1 represents the entire process of exploration and discovery.

The specifications in the exploration equations are consistent with what

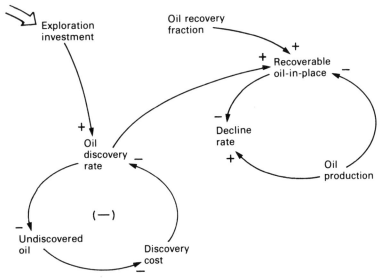

Figure 3.1 Exploration and discovery. Note that the negative sign enclosed by parentheses indicates a negative feedback loop.

is known of the discovery process in the petroleum industry, but there are data problems. First, the total magnitude of undiscovered oil is not known precisely, although it can be estimated roughly given current knowledge of Gulf oil fields. Second, discovery costs are also approximate. There is very little data on discovery costs, although we do know that they are very low over the time period at hand. Estimates for some Gulf sources place discovery costs at an even lower figure. Third, while the relationships between exploration investments and discovery costs are not ambiguous, the precise functional form is not well known. However, the time delay resulting in proven reserves is well documented.

The discovery of new oil-in-place tends to be unpredictable in any particular field. A field or reservoir of undiscovered oil is always finite, but the boundaries of the field are not known. As it is further developed, beyond a certain point the costs of finding greater quantities of oil-in-place increase. As noted in figure 3.1, the costs of discovery grow as the quantity of undiscovered oil or deposits declines. This relationship is expressed as a variable that simply compares the level of deposits (termed undiscovered oil) at a given moment to the initial level in 1970. That fraction is set as the initial quantity of undiscovered oil (500 billion barrels) minus the amount of oil remaining undiscovered, divided by that initial quantity. The functional relationship is specified in figure 3.2.

Since a reservoir is finite, as it is further developed, the probabilities of getting close to the edge increase. Risks grow. The decline rate is specified endogenously as a ratio of oil production to recoverable oil-in-place. The higher this ratio, the greater the decline rate. Without further exploration,

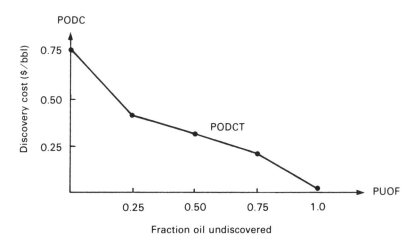

Figure 3.2 Functional relationship depicting oil discovery cost

there is an almost inevitable increase in costs. This fact entails additional built-in costs for petroleum production. As modeled, production costs are disaggregated into *discovery* costs associated with exploration, and *development* and operating costs.[4]

Production Capacity

The second stage in oil production is that of developing reserves. Investments in development produce the capital necessary to bring oil on-line. The quantity of capital required is related to production costs, the rate of capital depreciation, and the development investments (these relationships are described more fully in chapter 5). Figure 3.3 depicts the process of generating production capacity.

Production capacity is the rate of production (in barrels) from existing wells that could be maintained for about twelve months without additional development or marked loss in long-term recovery.[5] As modeled, production capacity is a direct function of investments, capacity costs, and depreciation. In 1972 Adelman estimated that the costs of producing one barrel of Gulf oil did not exceed $.10 and that this estimate was expected to hold for at least the next ten years. Subsequently he revised these estimates and argued that costs were considerably higher in the Gulf and would continue to rise.[6]

In the Gulf region development and operating costs are about $.25 per barrel, and according to Adelman they are anticipated to increase to $2.50 by 2000. While detailed explanations for such increases are not given, they are generally attributed to increases of capital costs, growing labor costs, and inflation.[7] Development costs are therefore specified as an increasing function to time, initialized at $.20 in 1970 and rising to $1.20 by 2000.[8] This trend refers to production cost per unit capacity.

Oil Production

The third stage in the process of oil production is the actual lifting of oil from the ground. The amount of oil produced depends on the total demand for Gulf oil. The exporters will produce the amount necessary to satisfy that demand—unless that demand is greater than the Gulf states' production capacity. Oil production is therefore constrained by capacity.

Total Gulf oil demand is the sum of the consumer countries Gulf oil import demand and the producers' oil demand. Thus the total demand is the sum of demand from the consumer states and the producer states.

Figure 3.4 depicts the process of actual oil production. Note the impact

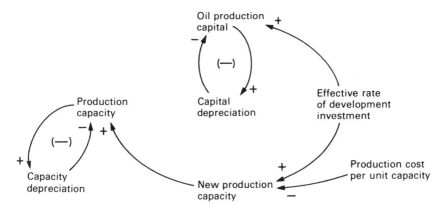

Figure 3.3 Production capacity

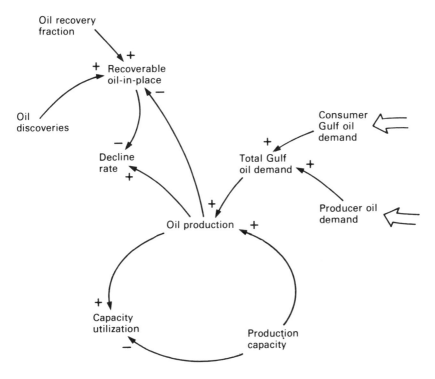

Figure 3.4 Oil production

of demand on production and the constraints imposed by production capacity.

3.2 Integrated Supply Sector

The three stages of the oil production process are represented in figure 3.5. By specifying this process in sufficient detail, it is possible to make explicit the type of influences exerted by each entity in international petroleum exchanges as decisions are made that affect the rate of oil production. This formulation then determines the impact of investments in exploration and development on production and production capacity and identifies constraints of production capacity upon capacity utilization.

The variables that are inputs into this sector of the model are the consumer countries' Gulf oil import demand, the producer countries' oil demand, the oil exploration investment, and the oil development investment. The demand variables constitute the requirements of importers and exporters for petroleum consumption. The investment variables represent the international oil companies' own contributions to the production of oil. The consumer Gulf oil import demand is computed in the consumer sector. The producer oil demand is computed in the producer sector.

The investment variables that are major inputs in the first two stages of oil production are set and calculated in the finance sector where the computations pertaining to financial flows are made. These two types of investments, once decided upon, then set in motion the production process presented in figure 3.4. The demand variables are summed to yield the total demand for oil from Gulf sources. That total demand is a major input variable determining the rate of petroleum production.

The output of this sector includes oil production capacity, oil capacity utilization, the decline rate, the rate of capital depreciation, and discovery costs. All these variables are then employed as inputs into the management sector where additional computations pertaining to the international oil companies' decisions are made. The supply sector also generates production available for export to the consumer countries. That latter variable is an input into the finance sector where it serves to compute the basis of expenditures in payment for oil imports.

The process of oil production, as modeled here, is highly dynamic. It is composed of a set of relationships depicted as linkages between absolute levels of oil and of capital, rates of change, and decision rules that cause the variables to change. The levels pertain to the reservoir of undiscovered oil, the recoverable oil-in-place (the proven reserves), and the un-

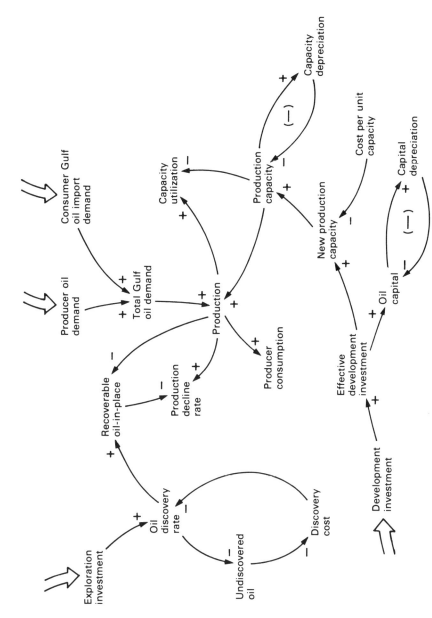

Figure 3.5 Causal loop diagram of the supply sector

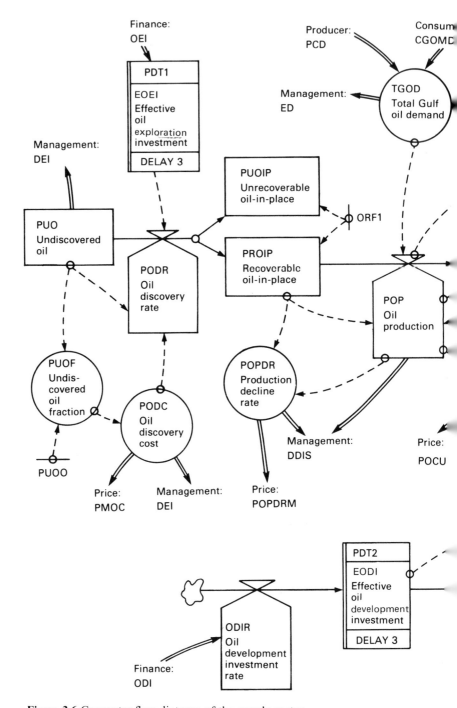

Figure 3.6 Computer flow diagram of the supply sector

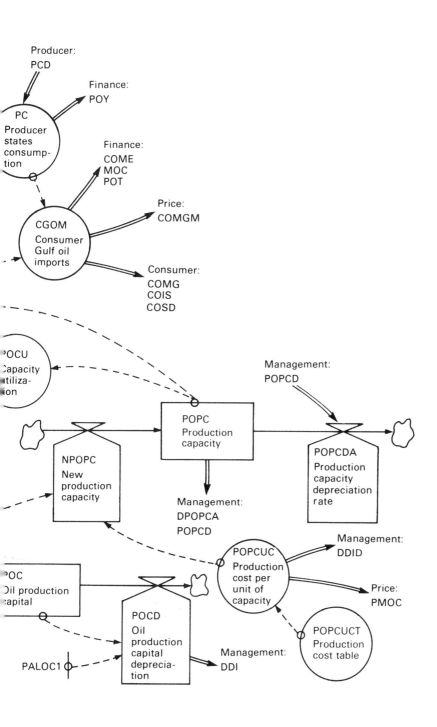

recoverable oil-in-place. These variables change with different rates of production, utilization, and decline. These rates are determined by the level of capital, which generates production capacity and the rate of depreciation of that capital. Underlying this entire process are the decisions made by the international oil companies and even more fundamentally those of the consumer and producer countries, as they make demands upon Gulf sources.[9]

The integrated computer flow diagram of the equations representing exploration, recovery, production, development investment, and cost is presented in figure 3.6. This diagram depicts the integrated structure of the supply sector and identifies the other sectors that influence and are influenced by supply-related variables. Appendix A lists the variables in the model for this and other sectors.

3.3 Linkages between the Supply Sector and the IPE Model

Oil production and the decisions governing the entire process are influenced by the total demand for Gulf oil and the total investments allocated for meeting that demand. The attempt to meet the demand for oil is made—whether it is in fact met or not is yet another matter. The consequences of meeting demand affect both the structure and the process of international petroleum exchanges.

First, the international oil companies are faced with a set of indicators that are employed in the calculation of investments for the next time period.

Second, an evaluation is made of the decline rate and the depreciation of oil capital. Both variables are stipulated as the concerns of the international oil corporations, along with investment, and used in the management sector.

Third, the amount of petroleum produced (and by definition sold to the consumer countries once the producers have met their own internal demand) is now available for computing the consumer countries' oil import expenditures, which can be computed once price is determined, as can the producer countries' oil income and the profits of the international oil companies. These three variables are calculated in the finance sector that provides general accounting functions for the international transactions around petroleum.

Fourth, these specifications involving the outcome of the production relationships at one iteration then contribute to the computation of expected demand for oil. (Expected demand is calculated in the management sector.)

Fifth, capacity utilization and the decline rate are used to estimate the companies' markup in the price sector. These variables are used by the corporations to influence the price of oil in response to the market situation according to their perception of investments that will be required in the oil industry.

4
The Finance Sector: Import Payments, Corporate Profits, and Oil Revenues

The finance sector is designed as the conceptual and computational basis for modeling corporate profits, consumer oil payments, and producer petroleum revenues. In this sector are characterized the financial interactions among the three entities in world petroleum exchanges, the implications of oil flows and dollar payments, and the process by which the companies allocate portions of their investments in exploration and development.

4.1 Financial Interdependence in Petroleum Trade

Consumer oil payments, producer revenue, and corporate profits are all predicated on unit price times quantity sold (or purchased). Producer oil revenues are determined by the tax rate per barrel and the number of barrels sold. Similarly consumer oil expenditures are determined by price times the quantity purchased. Corporate profits are based on oil payments of the consumer countries less corporate oil costs and producer oil taxes. Computationally, however, there are added intricacies in determining profits, payments, and revenues.

The calculations described in this chapter should not obscure the dynamic interrelationships and feedback links that reflect the interconnections throughout the model. The two major investment variables that have been treated as exogenous inputs into the oil production process, designated in the previous chapter, are endogenously generated in the finance sector. Consumer oil payments are obviously essential in calculating balance of payments and other critical variables, and petroleum revenue is a major input into the processes and decisions of the producer sector. Corporate profits are important in order to calculate investment in exploration and in development. To compute corporate profits, however, it is necessary to take into account the magnitude of the consumers' oil import expenditures and the producers' tax on the sale of petroleum.

This interdependence reflects the relationships created by the flow of petroleum across national boundaries. This chapter calculates (1) expenditures of the consumer countries in payments for petroleum imports, (2) income of the oil-exporting countries from oil sales, (3) corporate profits from oil trade, and (4) overall investments in the oil industry. These calculations all reflect dollar transactions.

Consumer Oil Import Expenditures

The payments of the oil-consuming countries are determined by two factors: the price of oil and the volume of imports. At each point in time payments are calculated as the number of barrels imported multiplied by price per barrel. Here too one can observe the leverage of the two agents in the oil market: the producer countries can modify the rate of payment, and the consumer countries can alter the amount imported. Of course the volume of consumer oil import expenditures appears as a debit against the consumer countries' balance of payments.

Producer Oil Income

The major source of producer revenues, termed income here, is taxes and royalties. The producer countries' own consumption of petroleum could in the long run provide claims on the total volume exported. In the past that consumption has been negligible, but it could change. Producer oil costs, which have also been negligible to date, may impinge on the magnitude of oil revenues over the longer run.

The tax rate of the producer countries, the major exogenous and policy variable in the model, represents the rate per barrel levied by the producer countries in their sales of oil. The historical tax rate—1970 to 1978—is used as the policy input in the reference analyses.

Corporate Oil Profits

It is extremely difficult to obtain an accurate view of the companies' profits. It is generally acknowledged that "profits and rates of return in the published financial statements have no clear or close resemblance to the rates of return that guide management decisions." [1] Nonetheless each year the companies are confronted with certain options in the development of the oil industry and in the allocations of investment, distribution of output, and regulation of flows. A seemingly valid argument would be that there is an underlying generic process that describes the management of the oil industry and that year-to-year deviations from that process reflect corporate responses to the characteristics of a particular situation.

For modeling purposes consumer oil payments constitute the only source of corporate oil profits. Given a certain markup, the more oil the consumer countries import, the greater will be corporate profits. But the multinational companies are also faced with built-in claims that are direct debits against profits. First, there are the taxes of the oil-producing countries, which presently constitute the largest component of price. Second, there are the corporations' own costs of oil production. Conceptually, corporate profits equal the quantity sold to consumers times the companies' share of price per barrel, namely, their markup. Computationally costs and taxes must be subtracted from total oil payments to yield corporate profits.

Despite the many uncertainties regarding the precise magnitude of corporate profits, it is possible to make some rough estimates of the division of gain from the sale of petroleum between the oil companies and the producer governments. The estimates presently available antedate the price increases of October 1973, when it was argued that the division of gain in mid-1969, once costs and taxes were taken into account, was about 83 : 17 in the producer countries' favor, ". . . a long way from 50–50." [2] This calculation is predicated on average f.o.b. receipts for Gulf oil of about $1.20, with total costs at about $.10. The tax therefore constitutes that portion of profits that accrued to the oil-producing countries. These historical figures are relevant here not because they have any direct bearing on price determination but because they indicate the relative gain for the oil companies and for the producer countries at the initial period of our simulations. The price increases of October 1973 were an intervention that occurred three years into the model's time frame. We can therefore observe their effects over both the long-term and the more immediate period. The subsequent tax increase changes the overall profits-revenue-payments relationships throughout the entire model. Figure 4.1 depicts these relationships.

Investments in Exploration and in Development
Once the price of petroleum and the quantity purchased are determined and corporate profits are calculated, then part of the profits is repatriated, and the remainder is reinvested in the petroleum industry of the oil-exporting countries.[3] Repatriated profits appear in the balance of payments position of the oil-importing countries. This process will be described in chapter 9. Reinvested profits are allocated to exploration and development.

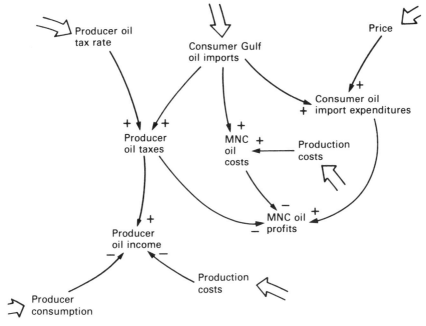

Figure 4.1 Financial flows from oil sales

For the industry to continue to produce over the long run, new reserves must be found and developed. How much exploration and development activity is actually undertaken depends upon current rates of production and expected demand, depreciation of existing production capital, depletion of reserves, costs of exploration and development, and time delays between the beginning of such undertakings and the actual bringing on-line of new reserves and new production equipment. Details are in chapter 5, which depicts the management sector of the IPE model.

In the basic model investment activities are financed out of corporate profits. Eighty percent of profits are considered to be available for reinvestment. This percentage refers to the maximum amount of profits the companies are willing to reinvest.[4] In the future the oil-producing countries may themselves seek to finance exploration and development as they consolidate their control over day-to-day operations. It is important to recognize the potential role of the producer countries in influencing the disposition of corporate profits. This influence enables them to maintain some control or leverage over the oil companies. The variable "available for reinvestment" is designed to represent these broad relationships.

In financing oil investments then, the companies will take account of the total desired amount of investments (based on calculations that we will describe later in the management sector) and then calculate the amount to be reinvested based on their profits for that year and the fraction specified to be available for reinvestment in the oil industry. They can invest from their profits as much as they need, given the production levels in the industry.

If reinvested profits do not cover the desired amount of investments, then the companies will draw more extensively from remaining profits—usually termed earned surplus—to finance investments.[5] So total oil investments are calculated as the sum of the corporations' reinvested profits, the producer countries' own investments, if there are any, and the direct oil investment which occurs when the reinvested profits are insufficient to meet desired investment targets in the industry.

In the basic simulations for this sector reinvested profits are always high enough to cover the corporations' desired investments. Is this a reflection of reality or an artifact of model specification? Information on this issue is extremely difficult to obtain. Validating our own estimates of corporate profits is close to impossible given corporate secrecy, transfer pricing, accounting procedures that do not designate what percentage of worldwide profits are earned in the Gulf, and so on. Nonetheless the conceptual basis of the investment variables and their importance to the nature of international petroleum exchanges cannot be denied.

The next question is: What difference does it make—conceptually and computationally—whether investments in exploration and development are financed from overall profit or from that fraction of profit (termed reinvested profits) assigned to the oil industry by the companies? The query is predicated of course on the consideration that both sources are based on the initial amount of profits obtained from the quantity of petroleum purchased by the consumer countries.

The answer is to be found in the underlying motivation for investment. All investment-related functions are designated with one end in mind, namely, to ensure that reserves and production capital are sufficient to permit oil production to meet demand. To date corporate profits have been more than sufficient to finance the amount of investment desired; countries and companies have not conflicted extensively on the nature or magnitudes of such investments, and the oil industry in the Gulf is not suffering from investment shortages. Under these conditions therefore the distinction between total corporate investments and that fraction reinvested in the oil industry is not significant. But it could be significant if changes occur in any of these three respects.

In summary investments in the oil industry are financed out of profits. The actual decisions that govern how much is invested and in what way are determined in the management sector. The factors that enter into the computation of oil investments indicate the multiple determinants of investments in the discovery of oil deposits and the development of reserves. Clearly the managers of petroleum exchanges make calculations as to desired levels of capacity utilization and oil production, and such calculations are central to the determination of investments in the oil industry overseas. Making financial allocations from corporate profits is only one step in that direction. The functional relationships designed to represent the financial basis of investments in exploration and in development are depicted in figure 4.2.

4.2 Integrated Finance Sector

Figure 4.3 depicts the processes that generate corporate profits, the relationship to consumer oil import expenditures and producer oil taxes and the connection with the companies' investments in the oil industry. Profits that are repatriated have balance of payment effects and appear as an entry in the current account.

Figure 4.4 presents the integrated computer flow diagram for the equa-

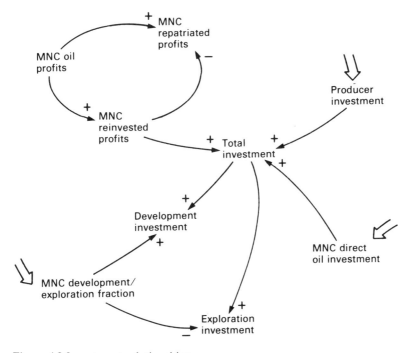

Figure 4.2 Investment relationships

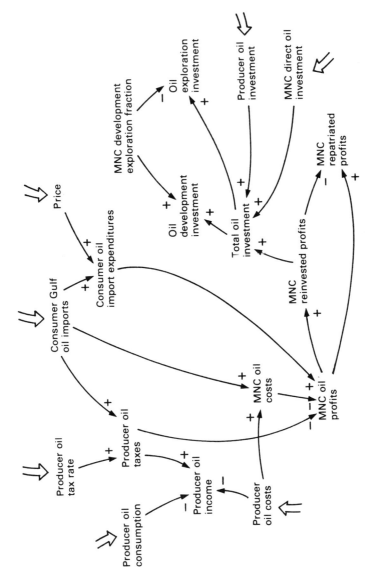

Figure 4.3 Causal loop diagram of the finance sector

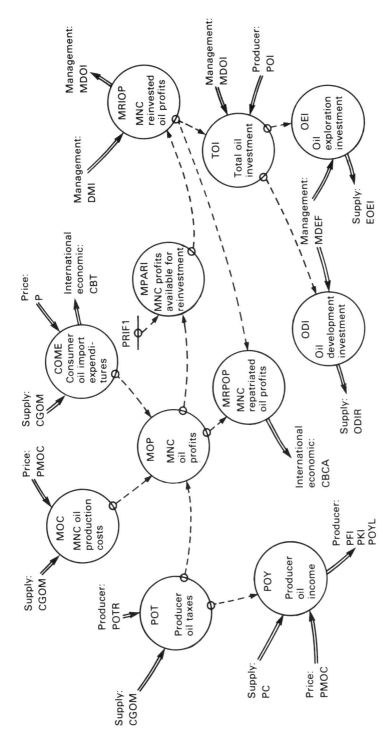

Figure 4.4 Computer flow diagram of the finance sector

tions in the finance sector, indicating linkages with other sectors and with the core of the model, namely, the supply sector.

4.3 Linkages between the Finance Sector and the IPE Model

The finance sector is designed basically as an interface module, providing critical inputs into the modeled oil production process and in turn the basis for the balance of payments calculations of the consumer countries. The calculations of consumer oil import expenditures reveal some of the economic implications of oil prices and amounts bought. The computations of the producer oil income indicate the effect of price times the amount sold once the corporate share of price is taken into account. The corporate profits are computed to generate the availability of funds for exploration investment and development.

The decisions governing investment allocations in the oil industry are made in the management sector. These decisions represent the managerial functions of the international oil companies, as they weigh the demands of the consumer and producer countries and the requirements of the oil industry itself. In this respect the managerial role of the companies and the structural context for controlling petroleum exchanges become apparent. The linkages of corporate profits to the overall exchanges in oil are those mapped out in figures 4.3 and 4.4. The figures indicate the ways in which profits are channeled to investments in oil exploration and development. Therein lies a major source of corporate control of the oil industry. The management sector reveals more clearly the processes of corporate regulation of an international industry.

5
The Management Sector:
Corporate Investments in the Oil Industry

The process of oil production described in chapter 3 begins with investment in exploration for deposits and investments in the development of reserves. Throughout the history of the international petroleum industry, these investments have been made by the multinational corporations. In the past the companies have controlled the pace of the expansion of the petroleum industry and petroleum trade. The governments of the oil-rich countries have only recently sought to direct management decisions. However, the oil companies continue to influence the development of petroleum reserves and the investments in exploration and development.

Although the degree of concentration in the oil industry has decreased over the past twenty years, it is still very high. The principal concessions of the petroleum companies in the Gulf area accounted for 26 percent of world production in 1966. By 1970 it was closer to 25 percent. On balance, market control by the international oil companies persists, and observers generally acknowledge that it is considerably greater than these figures would suggest.[1] The mechanism for attaining and perpetuating market control has been described extensively elsewhere.[2] Here it is useful to recall that sales contracts, joint ventures, and integration into the consumer market enabled each company to have sufficient information about the activities of every other oil company as to reinforce a system of mutual coproduction and cooperation.

This chapter depicts the process by which the oil companies allocate portions of their profits to investments in the oil industry for exploration and development. The companies make their investment decisions by taking into account the demands of the oil-importing countries, the supply situation, and the anticipated costs of production. These factors are central to the determination of exploration and development investments and to corporate markup on price. The essential point is not that corporations make exactly the decisions specified here but rather that these decisions can be formulated as general processes.

Various scenarios have been used to characterize the behavior and plans of the international oil companies. At least three of these views can be identified: first is the perspective dominant in many quarters in the West that the company interests are synonymous with those of the oil-exporting countries, that higher prices result in profit for both entities. Second is the view in developing countries that argues a convergence between the interests of the oil companies and the consumer countries. The third view is that of analysts who have a transactional view of the world and regard the oil companies as independent entities. The specifications embedded in this model are not designed to support any of these views but recognize the historical importance of the oil companies in shaping the oil market, the evolution of market control, and their continued importance despite initial changes in the role of the oil-exporting countries. The debates revolving around the corporations' control of the international oil industry involves not only manipulating investments, supply, and distribution but also price. The success of the oil-producing countries in raising the price of oil in October 1973, and since then intermittently to the present time, may overshadow the influence that the corporations have had, and continue to have, on price specification.

5.1 Investment Decisions

Investment decisions involve making assessments of the prevailing production and demand conditions in the oil industry, raising the required financing, and making appropriate allocations to exploration and development. The finances for oil investments are obtained by drawing upon corporate profits. Allocating investment funds to exploration activities and development entails some comparisons of actual with desired conditions. Exploration investments involve the comparison of the current reserves situation in relation to the desired decline rate in production, and development investments entail the comparisons of present production capacity to some anticipated production requirements.

The initial assessment of investment requirements are made in consideration of the desired decline rate in reserves (a critical factor in determining desired discoveries and, by extension, exploration requirements) and the desired capacity utilization (essential to determining investments in development).

Desired Exploration Investments

Conceptually the exploration investments modeled in this chapter begin with the stipulation that the oil companies seek to discover as much oil as they in fact produce when the decline rate is at the desired level and that the decisions governing exploration involve a comparison of actual and desired decline rates.[3] The decline rate is of course the rate at which reserves are depleted, given prevailing levels of oil production. If proven reserves are being depleted more rapidly than desired, then the companies will seek to discover more oil than the volume actually produced. Conversely, if reserves are being depleted at a rate lower than the oil companies consider appropriate, then the companies will seek to discover less oil than was produced. In the IPE model the desired decline rate in the Gulf is set as if it represented the concerns of the oil companies. In the late 1970s a new expression of concern for safeguarding natural resources led some producer governments to view the decline rate as one of the most important factors in the determination of petroleum policy.

In sum the oil companies' desired oil discoveries contribute to setting the desired exploration investment. The volume of reserves required determines the amounts of investments to be allocated toward the generation of reserves from the underlying deposits. The higher the level of oil production, the greater will be the desired discoveries. If the underlying reserves are drawn upon at a growing rate, the corporations will make the decision to increase the investments in exploration. The costs of discovery, however, are an important constraint on the desired exploration investments. Figure 5.1 depicts these relationships.

Determining the desired discoveries is the first step in assessing prevailing conditions. This assessment involves the interconnections between production and the decline rate, on the one hand, and target variables such as the desired decline rate, to generate what is considered to be an appropriate amount of desired discoveries, on the other. The differences between the actual and the target variables, and the costs of oil discovery, together contribute to the corporations' definition of the desired investment allocations.

Desired Development Investment

In calculating the desired amounts of investments to be allocated to development, we postulate that the international oil companies take into account two factors: production cost per unit capacity and the additional capacity desired. Production cost per unit capacity is estimated according to the specification described in chapter 3. Additions to capacity are based in turn on three considerations: existing production capacity, desired production capacity, which is derived from expected demand, and capacity loss incurred during the course of production. Capital depreciation is set at approximately 10 percent per year in the Gulf region. The

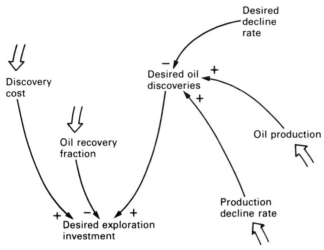

Figure 5.1 Desired exploration investment

costs of production capacity and the desired additions determine the companies' desired development investments in the oil industry. They will in turn be used to determine actual investments.

Oil Production Capacity Production capacity is a central feature of both the oil production process and the companies' calculations of their investment requirements. It is important to distinguish between actual and desired production capacity. Actual production capacity refers to the volume of oil that could be produced at a given period, and it depends on the amount of oil capital in place, unless proven reserves are at a level where depletion is a constraint. As modeled, production capacity is determined by the amount of previous production capacity, new investments and the cost per unit capacity, and capacity loss to depreciation (see chapter 3). Desired production capacity, by contrast, refers to the amount of capital required by the oil companies to obtain the level of capacity required to meet expected demand. This chapter is concerned solely with how the oil companies determine desired production capacity, and not with the actual production capacity. The desired capacity is determined by evaluating the overall conditions of the oil industry.

Desired Production Capacity Two factors serve as signals of the conditions in the oil industry and guide the companies' estimations of desired production capacity. First is the desired capacity utilization, which is set at 0.9 to represent the situation in which the multinational oil companies utilize 90 percent of the production capacity-in-place. The idle 10 percent is kept to allow for normal maintenance, for replacement in the event of unanticipated production problems, and for greater production in times of greater demand. Second is the expected demand for Gulf oil. This demand is modeled more fully in a subsequent chapter. Expected demand is a function of total demand for Gulf oil (from all sources of oil consumption) and the consumer countries' import gap (the difference between consumer demand from the Gulf and actual imports). The higher the expected demand for oil and the lower the desired capacity utilization, the higher will be the corporations' desired production capacity. With greater requirements for production capacity, there will be more desired additional capacity. Conceptually desired oil production capacity functions as a guideline for investments in the oil industry. Computationally this guideline depends on prevailing oil demand and the utilization of existing capacity.

Desired Additional Capacity   Although the multinational oil corporations seek to expand capacity as long as demand sustains such an expansion, if the volume of new capacity desired is so low that it is less than the amount of depreciations on oil capital already in place, then capacity will be allowed to slowly shrink by attrition. Only 80 percent of that which depreciated is replaced by added investments.

The additional capacity required by the oil companies will therefore be the difference between desired capacity (in bbl/year), on the one hand, and production capacity (in bbl/year) minus capacity loss through capital depreciation (in bbl/year), on the other.

Figure 5.2 characterizes the impact of expected demand upon required oil production capacity and desired additional production capacity. These linkages represent one means by which the international oil companies seek to meet the demands of the consumer countries in relation to the available capacity. They also indicate the ways in which the companies could induce shortages in supply to maintain fuller control of international petroleum exchanges. By knowing the expected demand and by tailoring production capacity to (presumably) meet these specifications, the oil companies could create a tight supply situation almost at will. Shortages of supply could be induced by planning for relatively less production capacity than would be warranted, given expected demand— alternatively, a glut of oil could be produced by increasing additional production capacity. Such manipulations would invariably have an impact on price. Excess capacity would depress markup and contribute to the inflation of production costs; insufficient capacity would increase markup with no reduction in production costs. Thus the companies' calculations of desired production capacity and additional capacity will influence oil prices.

Capacity utilization assists in the determination of development invest-

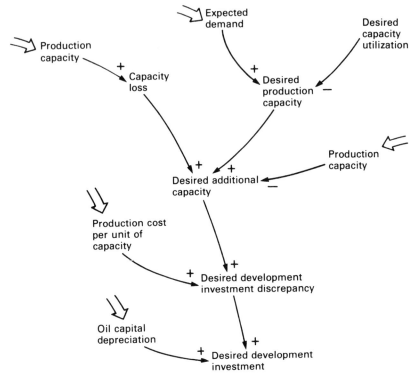

Figure 5.2 Desired development investment

ments and provides one important input into the markup on price. The product of desired additional capacity and cost per unit capacity generates the desired development investment discrepancy. Computationally desired development investment is set equal to development investment discrepancy as long as that discrepancy is greater than the amount of capital depreciation. If depreciation is greater than the discrepancy, the desired amount will be set at 80 percent of the amount depreciated.

Direct Oil Investment
Oil companies make direct investments based on their calculations of shortfalls in investments by the producers and their own profit reinvestments. This discrepancy results in the allocations from the oil companies' working capital to the oil industry. Computationally we set the producers' investments in the oil industry to zero in the reference simulation, reflecting the limited role of the governments in 1970 when oil investments were made and controlled by the companies. These specifications are presented in figure 5.3.

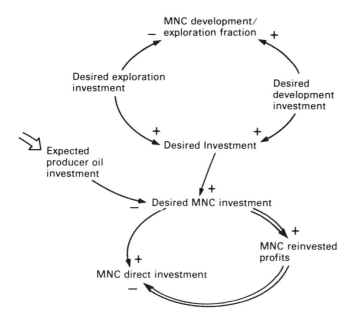

Figure 5.3 New oil investment. Double lines indicate intersector flows.

5.2 Integrated Management Sector

The conditions of the oil industry provide important information for the corporations' determination of the investments required for the Gulf region. These calculations are based on information about oil production, the decline rate and the discrepancies between what is actually happening and what the corporations' desired targets are. This view of goal-seeking countries and companies is pervasive in investment specification. The difference between actual and desired steers corporate behavior and generates outcome.

Changes in this sector can easily be made to reflect corporate decisions. For instance, desired capacity utilization, now set at 90 percent of actual capacity, could be shifted, if corporations' preferences change. Also the basic markup set at the 1970 level of markup on Saudi light crude could be adjusted to reflect different assumptions. The impact of corporate strategies on price is a critical aspect of this analysis. Despite the role of the producer countries in setting the tax rate, the companies make an important contribution to the final price. Debates about the extent of that contribution and the prevalence of a symbiotic rather than a competitive relationship between oil companies and producer countries are not uncommon.[4]

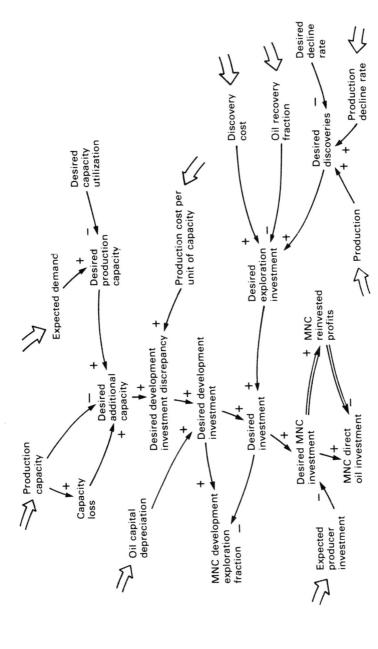

Figure 5.4 Causal loop diagram of the management sector. Double lines indicate intersector flows.

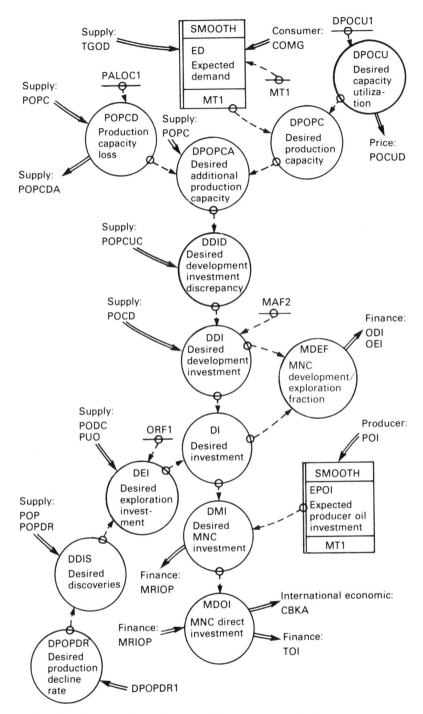

Figure 5.5 Computer flow diagram of the management sector

Figure 5.4 depicts all the linkages in the management sector. Figure 5.5 presents its computational structure. The investment relationships in this sector yield information central to the computations of other sectors and to the generation of endogenously determined behavior when the IPE model is integrated.[5] Indicated are the investment allocations of the international oil companies and the calculations they make as they direct fractions of their profits toward support of the oil industry in the Gulf region. In each case the conditions prevailing in the oil industry provide a context juxtaposed against the corporations' preferences, and the outcome is in the form of a markup.

5.3 Linkages between the Management Sector and the IPE Model

The major inputs in the formulations of the management sector are derived, first, from specifications of oil production in the supply sector which pertain largely to actual production, including production capacity and the decline rate. These inputs are the signals oil companies use in calculating their desired investments in the oil industry. The second set of inputs are provided from computations of oil import demand in the consumer sector, largely the demand for imports from the Gulf area. The third set of inputs is from the producer sector where oil income could be used in the future to determine the producer countries' contributions to the investments in the oil industry.

Once the calculations of management are made concerning the oil companies' investments in exploration and development and their oil production capacity, price can be determined.

6
The Price Sector

The IPE model acknowledges the special role of the oil-producing countries in setting the tax rate, and also the conditions of oil production and the corporations' determination of markup on price. Each component of price is generated in other sectors. In this chapter the calculations for the final determination of price are presented. The price computations integrate the model as a whole.

6.1 Price Determination

The price of crude petroleum is specified as a function of the producer countries' tax rate per barrel of oil, the corporations' markup, and the cost of production.[1]

The tax rate is set exogenously to the model as a whole and specified in the producer sector. It is an important policy variable in the model, since experimentation with alternative taxes enables the determination of systemwide implications for all entities in international petroleum exchanges.

In 1970 production cost was a relatively marginal component of price. In the Gulf region costs were largely negligible. In other areas of the world production costs are more substantial. Everywhere, however, discovery cost and production cost per unit capacity contribute to the final determination of the cost of petroleum production. Over time, in different areas and with different rates, production costs will increase.

The markup on price is determined and controlled by the international oil companies. In essence markup represents the convergence of supply and demand forces that influence corporate decisions and the corporations' efforts to make adjustments in supply and demand. This is not to suggest that all supply and demand interactions are captured by markup but that the most causally proximate influence of supply and demand forces upon price is through the companies' markup.

Markup is influenced by the decline rate, the degree of capacity utilization, and the difference between consumer Gulf oil demand and actual oil imports. This differentiation among the factors that potentially effect a shortfall of imports—embargo possibilities (bearing on imports directly), or depletion possibilities (bearing on production and reserves)—will provide insights into various impacts upon markup, price, and balance of payments variables. By observing the supply and demand forces in the world petroleum market, the international oil companies are then able to adjust markup accordingly. A more precise description of the markup factor—and its underlying process—will reflect its inherent dynamic impact upon price. Indeed the only dynamic feature of the price specification is incorporated in markup. The other components of price are either exogenous (the tax rate) or relatively stable in the short run (production costs).

The representation of price is presented in figure 6.1. The formulation is strictly an aggregative one. The determination of the individual components of price represent the dynamic relationships of the model.

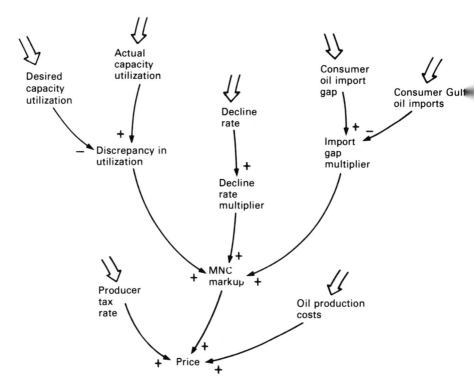

Figure 6.1 Causal loop diagram of the price sector

The Tax Rate

The producer countries' tax rate per barrel is an essential component of price. However, the other components cannot be ignored, for they may have substantial effects on the final determination of price. Depending on production rates and depletion possibilities, for example, costs will increase. The base simulation—the reference case—of the model employs the historical tax rate, from 1970 to 1978, and then maintains the 1978 rate constant to 2000. Against this rate one can compare the effects of alternative tax rates on the model as a whole. These alternative specifications, and the results, are presented in part III.

Production Costs

The cost component of price is composed of two variables: oil discovery costs and production cost per unit of capacity. Discovery costs are formulated as a function of undiscovered oil, in that the discovery costs increase as the remaining fraction of undiscovered oil declines. These costs are not time dependent; they vary from $.02 in 1970 to $.75 when depletion occurs. Production costs grow from $.20 in 1970 to $1.20 in 2000. The rationale for these two production cost specifications are presented in chapter 3. Costs depend upon how much oil is actually produced, the period of time when it is produced, and how much oil is left in the pool of undiscovered oil. The cost calculations have been adjusted by $.10 in 1970 due to discrepancies among cost estimates.

Markup on Price

In determining markup, the oil companies begin with a base markup that corresponds to the 1970 per barrel markup on Saudi light crude, calculated by subtracting the 1970 costs and the tax rate from the 1970 price. This base is not a minimum; it is an historical figure employed solely for initialization purposes. Three other inputs on markup reflect the corporations' evaluations of their ability of meet consumer oil demand, the existing production capacity utilization in the Gulf area, and the prevailing decline rates.

The first input is the oil capacity utilization discrepancy. That variable is endogenous to the model and reveals the difference between actual capacity utilization and the companies' specification of the desired capacity utilization. As noted in chapter 4, capacity utilization is determined by the ratio of oil produced to presently available capacity. Desired capacity utilization, by contrast, refers to the corporations' own a priori formulation of what they consider as an appropriate level of capacity utilization. This consideration is of major importance, since much of what is publicly known about the oil industry overseas points to frequent attempts by the oil companies to reduce output so as to strengthen price—long before the oil-producing countries adopted that policy as a potential leverage in support of their own price structure.

Desired capacity utilization is set at 90 percent of capacity. It would be informative to vary that coefficient and observe its implications for the corporations' markup and, by extension, the attendant impact upon price. Such a large fraction is based on the cost of excess capacity with the need to allow for maintenance, interruptions, and unexpected increases in demand. The desired capacity utilization discrepancy therefore is the difference between the actual capacity that is employed and that which the companies view as appropriate at a particular point, and it serves as an indicator of the market tightness and need for investment in production capacity. That quotient constitutes the first endogenous component of markup.

The second component of markup is the decline rate in the Gulf, which is determined endogenously as the ratio of oil production to recoverable oil-in-place. The decline rate is incorporated in the markup specification as a multiplier which impacts upon the base markup. Thus the higher the decline rate, the greater will be the impact upon markup.

The third factor is the consumers' Gulf oil import gap. Computationally it is the difference between consumer Gulf oil import demand and actual imports from the Gulf divided by imports from the Gulf. The greater the difference between the Gulf supplies and consumer Gulf import demand, the stronger will be the impact on markup. The gap variable reflects the tightness of supplies. This factor too is treated as a multiplier upon markup.

The logic underlying this modeled specification of markup is partly based on our reading of the corporations' assessments and is partly an abstraction for computational ease to differentiate those factors generated endogenously and those essentially given. Computationally the base markup—set at $.82 for 1970—is given and is the initial point of departure for the corporations' calculation of markup. Then the capacity utilization discrepancy is added to the base price. Once this is done, that sum and the two other variables contribute multiplicatively to generate the corporations' final markup.

The price specification reflects the noncompetitive elements of the market generated by the strong producer influence on price and by corporate strength. At the same time there are competitive influences represented by the multipliers on markup. Again for a specification of the final price to the consumers in terms of end uses, a consumer tax rate would have to be incorporated in the price equations.

Figure 6.2 presents the computer flow diagram of the equations for the determination of price. Note the labels indicating linkages to other sectors of the model.

6.2 Linkages between the Price Sector and the IPE Model

Price determination is a central feature of the model in both conceptual and computational terms. The inputs into price come from the producer sector (the tax rate), from the supply sector (the cost components and the multiplier components of markup), and from the consumer sector (the oil import gap). Price has a critical role in the consumer sector, regulating within this sector supply and demand through the price elasticities as well

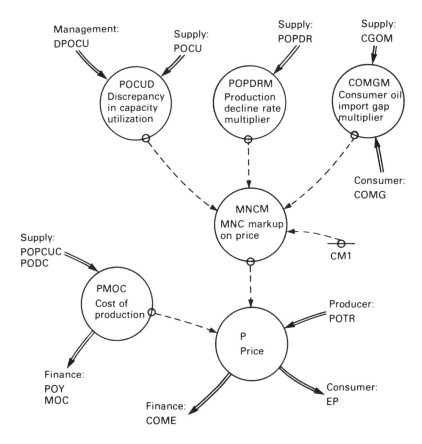

Figure 6.2 Computer flow diagram of the price sector

as influencing the availability of substitute energies. Once the price is calculated, then it is an input into the finance sector (as an essential aspect of the consumer oil import expenditures). The price effects are therefore transmitted throughout the model in terms of their impacts on the balance of payments, on the oil revenues of the producer countries, on their development and foreign investments and on the amount of oil bought and sold. The magnitude of oil trade determines the expansion of production and the existing reserves. These concerns in turn influence the investments in development and exploration. In the longer run the discovery and production per unit capacity costs will go up.

The essential features of the price specification of this simulation model are that the determinants and the consequences of oil prices are specified in the same conceptual and computational framework. It is therefore possible to identify the sources of changes in prices and the different effects of price changes as they are transmitted throughout the entire structure of international petroleum exchanges.

7
The Producer Sector:
Growth and Government Decisions

The purpose of this chapter is to determine the impact upon international financial transactions of the developments in oil-exporting countries and to describe and characterize the decisions of the governments as they attempt to manage their own development. The first objective entails modeling structural processes of economic growth, and the second, governmental interventions designed to control and regulate these processes. In particular this chapter looks at governmental interventions in domestic growth. Clearly the tax rate is not the only leverage available to the exporting countries nor the only means by which they influence worldwide petroleum exchanges. Furthermore it is not the only major decision confronting the governments of oil-exporting states.

7.1 Modeling Growth

The producer sector is designed not to represent the process of economic development in its entirety, or even to approximate its fundamental features. The focus is exclusively on three sets of factors: (1) the process generating the producer countries' domestic capital investments, since such investments constitute a fundamental indicator of industrialization and economic growth, (2) the demand of these countries for oil, since oil consumption is related to the extent of industrialization, which exerts in turn some claims upon available petroleum supplies, and (3) the producer countries' demand for imports of goods and services, since such imports constitute a direct input into the economies of the consumer nations and are an important means of redressing their balance of payments difficulties.

There is no average oil-producing country. It is, however, possible to delineate generic features of this entity in the world petroleum exchange.

Data for three fundamentally different countries—Saudi Arabia, Iran, and Kuwait—were pooled over a twelve-year period, 1962 to 1973; then the coefficients were estimated, and the results employed to generalize to other oil-exporting countries.[1]

The specifications in the producer sector are designed to clarify the claims these countries make upon petroleum supplies, their economic and financial relationships with the oil-importing countries, given increased oil revenues and the possibilities for increasing allocations to investments in the oil industry. The allocation of oil revenues and investments in foreign economies have an impact on the international economy, particularly on the balance of payments of the oil-consuming countries. Internal distributional issues are not examined in this chapter, nor are sectoral investments disaggregated or the new preferences for economic diversification taken into account. Rather, our consideration of the producer governments looks more extensively at domestic capital investments, domestic oil consumption, and demand for imports of goods and services from oil-consuming countries than do current modeling practices.[2]

7.2 Domestic Capital Investments

The domestic investments of the oil-exporting countries are determined by the availability of oil revenue, population size and capital stock. Income from the sale of petroleum is allocated to capital investments, and at the same time there is a tendency to increase the demand for such investments. Investment demand is not determined by oil income alone. The size of the population is a critical factor in shaping this demand. Population is conventionally used as a surrogate for a variety of underlying sociological and economic determinants of domestic investment that reflect market size. It is also necessary to take into account the rate of depreciation of capital stock, since the stock of capital will be depleted unless replacement investments are made. These relationships are presented in figure 7.1.

In figure 7.1 one positive feedback loop generates the oil-exporting countries' capital investments and by extension increases their capital stock; the other, negative loop, represents the depletion of that stock. Over time these relationships determine the countries' level of industrialization. To estimate these linkages, it is necessary first to specify the equations, second, obtain data on each variable, and, third, apply appropriate estimation procedures. Statistical methods used are discussed in a subsequent section of this chapter.

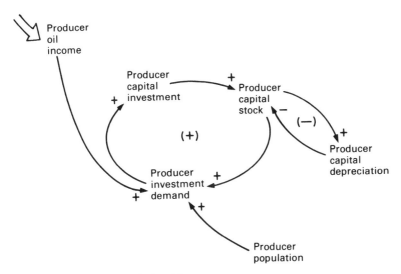

Figure 7.1 Investment relationships

7.3 Demand for Imports of Goods and Services

Import demand is stipulated as a function of oil income, capital stock (as an indicator of overall industrial levels), and the size of the population. The larger these imports, the more favorable the impact is for the consumer countries. Figure 7.2 depicts the oil-exporting countries' demand for imports of goods and services and their own demand for oil.

7.4 Demand for Oil

The oil demand of the exporters will grow as their level of industrialization increases and as they make larger investments in their own economies. As presently modeled, the demand for oil is determined by the producers' population and by their capital stock. Both variables generate a positive impact on the demand for petroleum.

Population is initialized at the 1970 level for the Gulf producers—at 47 million—and stipulated to increase at a rate of 3.1 percent per year, which is a weighted average of the population growth rates for those countries, as reported by a U.N. statistics survey.[3] The other determinant of the producers' oil demand, capital stock, is derived from data on gross domestic product.

Figure 7.3 presents an integrated diagram of the linkages in the producer sector. The major driving factor is the income from petroleum sales. Producer oil revenue is calculated in the finance sector and serves as a

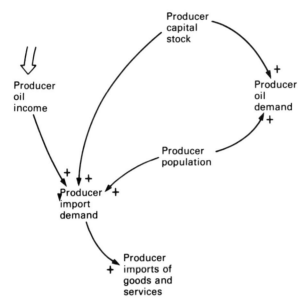

Figure 7.2 Demand for imports and domestic oil consumption

major input into the computations of the producer sector. Imports of
goods and services depicted on the left of figure 7.3 serves as a direct input into the consumer countries' trade balance calculations. The links between the producers' own usages of their oil revenue and the implications
for the consumer countries can be gleaned.

7.5 From Functional Relationships to Statistical Equations
in Modeling Growth

The relationships formulated in the producer sector represent the major
developmental features of the oil-exporting countries that have worldwide implications for petroleum exchanges. The linkages depicted in
figures 7.1 to 7.3 are reformulated as a set of statistical equations. The
equations are designed to yield empirical estimation of the determinants
of producer countries' domestic capital investment demand, import demand, and demand for oil.

Demand for Capital Investment
The producer countries' capital investment demand is estimated in terms
of gross fixed capital formation in non-oil sectors, since investments in the
oil sectors are undertaken by the companies.[4] But, since oil income ex-

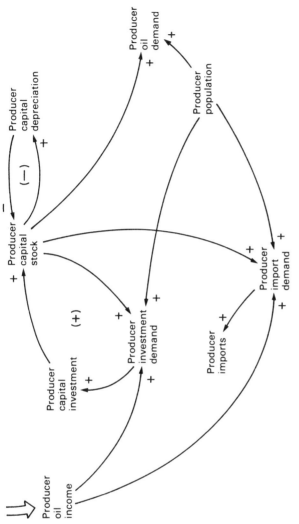

Figure 7.3 Causal loop diagram of structural relationships of the producer sector

ceeds absorptive capacity in these countries, it is necessary to base the investment relation on an estimate of absorptive capacity.[5] For the latter we employ non-oil investment demand, Y_1, specified as

$$Y_1 = \beta_{11}X_1 + \beta_{12}X_2,$$

where X_1 is gross domestic capital stock and X_2 is oil revenue. Then the relations are transformed into per capita terms (population, X_3) and a constant term (α) is included to obtain a nonhomogenous linear approximation, yielding

$$\frac{Y_1}{X_3} = \frac{\alpha_1}{X_3} + \beta_{11}\frac{X_1}{X_3} + \beta_{12}\frac{X_2}{X_3} + \mu,$$

which becomes

$$Y_1 = \alpha_1 + \beta_{11}X_1 + \beta_{12}X_2 + \beta_{13}X_3 + \mu_1. \tag{7.1}$$

Capital stock formation is then specified as a function of the level of capital stock in the oil-exporting countries, the rate of capital investment, and the rate of capital depreciation. The initialization of producer capital stock at \$50 billion is a rough figure for a key variable whose precise magnitude is essentially unknown. It is based on a common approximation used by development economists which sets the capital stock of a developing country at approximately 2.5 times the gross domestic product. The GDP for the six Gulf states in 1970 is summed and multiplied by 2.5 to obtain a rough figure for capital stock. The rate of domestic capital investment in the oil-exporting countries is the essential input in the formation of producer capital stock.

Producer Demand for Imports
The rationale for import demand is different. It must be noted initially that import demand (Y_2) is for two kinds of goods, consumer goods and investment goods. Imports of investment goods are

$$Y_2^I = \gamma(\alpha_1 + \beta_{11}X_1 + \beta_{12}X_2 + \beta_{13}X_3),$$

where X_1, X_2, and X_3 are defined as in equation (7.1). Imports of consumption goods are related to gross domestic product:

$$Y_2^C = \delta X_1.$$

Thus total imports demand is

$$\begin{aligned} Y_2 &= Y_2^I + Y_2^C \\ &= \alpha_2 + \beta_{21}X_1 + \beta_{22}X_2 + \beta_{23}X_3 + \mu_2, \end{aligned} \tag{7.2}$$

where the coefficients are

$$\alpha_2 = \gamma\alpha_1,$$
$$\beta_{21} = \gamma\beta_{11} + \delta,$$
$$\beta_{22} = \gamma\beta_{12},$$
$$\beta_{23} = \gamma\beta_{13}.$$

The producer countries' import demand equation does not represent actual imports. A distinction is made between all import requirements met and the remaining unmet demand. The base simulation analysis stipulates that import demand is entirely met. However, when this issue is treated as a government decision, and the effects of that decision are simulated, a discrepancy between the demand for imports and actual imports can occur. A country's imports may be constrained through the inability to meet its payments, through domestic constraints such as those of absorptive capacity, or even as a result of conscious government policy. To date the major developmental constraint in oil-exporting countries has been generating the demand for imports rather than meeting that demand. Imports bear directly upon the consumer countries' balance of payments, representing an individual entry against the current trade balance. The producers' import demand is thus another major linkage variable, connecting producers and consumers and serving to integrate different sectors of the model.

Producer Oil Demand
The demand for domestic oil consumption (Y_3) for the producer countries is specified as a function of population (X_3) and capital stock (X_1). Thus

$$\frac{Y_3}{X_3} = \frac{\alpha_3}{X_3} + \beta_{31}\frac{X_1}{X_3},$$
$$Y_3 = \alpha_3 + \beta_{31}X_1 + \beta_{33}X_3 + \mu_3 \tag{7.3}$$

where per capita capital stock represents both consumption and nonconsumption demands.

Recall that the oil demand specification for the IPE model as a whole is in two parts: the consumer countries' demand for petroleum (determined in the consumer sector, chapter 8) and the oil-exporting countries' own petroleum requirements, which are determined in this sector. Over time the growth of these countries generates increasing claims against petroleum supplies, and it is therefore important to model the process by which such claims are generated.

7.6   Statistical Formulation

The coefficients for the three development equations (7.1) to (7.3) of the producer sector are estimated from empirical data. We now look in more detail at the coefficients associated with these equations:

$$Y_1 = \alpha_1 + \beta_{11}X_1 + \beta_{12}X_2 + \beta_{13}X_3 + \mu_1,$$
$$Y_2 = \alpha_2 + \beta_{21}X_1 + \beta_{22}X_2 + \beta_{23}X_3 + \mu_2,$$
$$Y_3 = \alpha_3 + \beta_{31}X_1 + \beta_{33}X_3 + \mu_3,$$

where

Y_1 = gross fixed capital formation,
Y_2 = imports of consumption and investment goods,
Y_3 = oil consumption,
X_1 = gross domestic product,
X_2 = oil revenue,
X_3 = population,

and, the coefficients,

$\alpha_1 \ldots \alpha_3$ = intercept or constant,
$\beta_{11} \ldots \beta_{32}$ = regression coefficients,

with

$\mu_1 \ldots \mu_3$ = errors or disturbances.

The intercept term α is included to reduce specification errors inherent in a linear approximation of a nonlinear function. Data for the Y and X terms are obtained from the *International Financial Statistics* of the International Monetary Fund, the OPEC *Annual Statistical Bulletin,* and the *Petroleum Economist.*

The producer sector is specified as a generic producer with the aggregate combined characteristics of three Gulf states. Iran has a large population, lower oil reserves relative to other Gulf states, and a comparatively greater demand for oil income. Saudi Arabia and Kuwait both have small populations and considerably lower domestic claims against oil revenues. They differ only to the extent that Saudi Arabia has considerably larger known reserves than Kuwait, and its depletion horizon is therefore much longer. Employing these country-specific data to provide estimates for a generic producer assumes the absence of an undue distortion of known reality.

The paucity of data, the problem of multicollinearity, and the small number of observations make it difficult to estimate these equations country by country. The use of an alternative procedure for estimating the coefficients in equations (7.1) to (7.3) is necessary. This procedure essentially involves pooling data for the three countries and obtaining coefficient estimates from the pooled observations. Estimation is based on ordinary least squares calculation of the coefficients by transforming the raw data to purge the effects of autoregression (time-wise effects) and heteroscedasticity (cross-sectional effects).[6]

7.7 Results of Empirical Estimation

The coefficient estimates from the pooled observations and t statistics, noted in parentheses, are as follows:

$$Y_1 = -0.598 + 0.145X_1 + 0.104X_2 + 32.96X_3 + \mu_1, \qquad (7.4)$$
$$ (-2.15) \quad (8.64) \qquad (2.65) \qquad (3.53)$$
$$R^2 = 0.90,$$
$$F_{3,29} = 87.49;$$

$$Y_2 = 1.24 + 0.075X_1 + 0.045X_2 + 7.67X_3 + \mu_2, \qquad (7.5)$$
$$ (3.47) \quad (3.67) \qquad (0.53) \qquad (1.07)$$
$$R^2 = 0.89,$$
$$F_{3,29} = 76.01;$$

$$Y_3 = -0.35 + 0.0012X_1 + 1.95X_3 + \mu_3, \qquad (7.6)$$
$$ (-1.35) \quad (5.47) \qquad (13.91)$$
$$R^2 = 0.94,$$
$$F_{2,27} = 216.17.$$

In the simulation model the coefficients for the GDP variable, X_1, are assigned values 0.4 times those of their equivalent regression estimates. This adjustment is necessary due to the relationship of capital stock to GDP.

The coefficients for the main equations in the producer sector provide some basis for the computer specifications of this sector. There are further complexities, however. First, there might be cross-sectoral (or spatial) correlation in that what happens in Kuwait could affect Saudi Arabia and is not captured by the autocorrelation parameter. Second, the equations for imports, investment, and oil consumption have been estimated sequentially. It is possible that errors in one equation may be correlated with errors in another. A more complete specification of these relationships would undoubtedly involve a simultaneous estimation procedure. Since GDP is employed as a rough approximation of capital stock (with appro-

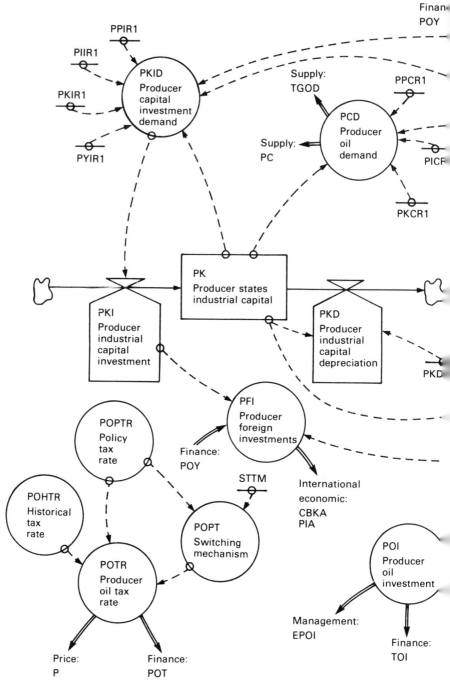

Figure 7.4 Computer flow diagram of the producer sector

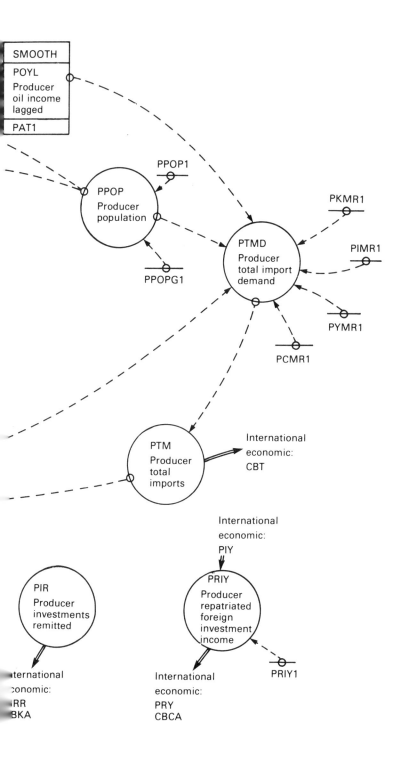

priate adjustment) and affects the demand for investment, these three equations represent only a partially specified system. Despite these potential difficulties the coefficients obtained by pooling cross-sectional and time series data have provided the producer sector with an empirically based set of estimates for the major coefficients. These estimates constitute a good approximation of the type and strength of the linkages in this sector.

7.8 Modeling Government Decisions

The major decision for producer countries is of course setting the tax rate for the sale of petroleum. Once that decision has been made, the magnitude of oil income can be determined. We do not model how that decision is made but look at the consequences of making that decision. Subsequent decisions, which are modeled here, have a direct implication for internal growth and for international economic relations.

The decisions bearing upon internal development are critical to determining how much of the domestic demand for capital investment is to be met and the extent to which demand for imports of goods and services will in fact be satisfied.[7] While we cannot determine precisely how or why these decisions are made, it is reasonable to assume that for the oil-exporting countries of the Gulf region all the demand for domestic capital investment will be met and that the governments will indeed satisfy the entire demand for imports of goods and services.

The economic situation of the oil-exporting countries is a policymaker's dream. Gulf countries are determining their own development via revenue from oil exports which support fiscal policy in the form of government programs. For example, the Saudis appear to be manipulating such demand to promote an economic growth rate that will not transform traditional power relations within Saudi Arabia while substantially increasing Saudi influence internationally. Kuwait and the other states in the Gulf region are confronted with a situation similar to that of Saudi Arabia, where generating investment demand is a major constraint to development. For Iran the shah's government set in motion an investment pattern that in time outstripped its capacity for making the appropriate financial allocations to support its commitments. All countries of the Gulf region are confronted with marked difficulties of absorptive capacity, their governments are aware of these difficulties, and there is some attempt to control the internal rates and patterns of growth and manipulate their direction by government policy.[8]

Three decisions that have direct implications for the oil-exporting countries' international economic relationships are particularly relevant: (1) the extent of investments in the economies of the oil-consuming countries, (2) the volume of oil income reinvested externally, and (3) by extension, how much income from investments is to be repatriated.[9]

In describing these decisions, it is reasonable to estimate that the oil-exporting countries' investments overseas will be determined by the difference between their oil income and domestically produced investment goods plus that spent on imports of goods and services of all kinds. The residual constitutes the total that will be held in foreign economies or held in gold reserves. Of the total holdings abroad a part will be directed toward long- as compared to short-term investments.

By the same token it is stipulated that at the present time all foreign investments remain reinvested externally, and that there is no short-term demand for repatriation of invested income. This simplification is entirely consistent with current realities since the problem facing the oil-exporting countries is one of surplus revenue. There is no reason for the producers to repatriate investment income as long as their domestic demand for capital investments and imports can be financed out of current income. In the years to come, that situation may change. The change can easily be specified in the repatriation equation.[10]

The computer diagram for the entire producer sector—including the growth process equations and governmental decisions—is presented in figure 7.4. The decisions are labeled in the lower part of the diagram, and the processes of growth and development in the center.

7.9   Linkages between the Producer Sector and the IPE Model

The major input into the producer sector is derived from the finance sector that computes producer oil income. Calculations are made on the basis of the volume of petroleum exported times the tax rate, less the amount of oil consumed by the producer countries themselves and the costs related to oil production. Once oil income is generated, the producer sector's function is to make use of that income for domestic investments, capital formation, and imports of goods and services.

In sum there are two principal outputs from the producer sector. One pertains to the oil-exporting countries' bill for imports from the consumer countries. The other involves the producers' investments overseas which also emerge in the consumer countries' balance of payments. These two processes reflect the essence of the recycling of oil revenue in the economies of the oil-importing countries.

8

The Consumer Sector:
Oil Imports and Strategic Vulnerability

The processes that dominate consumer behavior and contribute to the generation of oil imports from the Gulf region and the strategic vulnerability of consumer countries associated with oil imports are two issues central to any political analysis of international petroleum exchanges. The consumer countries' response to the price increases of October 1973 has been highly political in nature, and much of the ensuing reaction expressed concern for the apparent vulnerability associated with increased reliance on oil imports and higher oil prices. The economic processes modeled here should not obscure their political implications or the politics embedded in the formulation of import demand and strategic vulnerability for consumer countries.

Imports from the Gulf are specified as a function of four separate factors: demand for petroleum, oil production from domestic sources, total imports, and imports from the Gulf region. This disaggregation is an important characteristic of the IPE model, and one that represents a notable extension from prevailing modeling efforts.[1]

Once the demand for imports is calculated, we ask: What are the impacts of oil demand for the consumers' own strategic vulnerability? In answering this question quantitatively, we do not model directly the vulnerability in importing petroleum but the variables that enter the governments' decisions due to their perceptions about their vulnerability. Thus a formal representation of vulnerability is developed as a monitoring device that reveals the effects of different oil import policies.

Consumer vulnerability resulting from oil imports is based on four variables: (1) essentiality of petroleum to consumer economies, (2) insecurity of the sources from which oil is imported, (3) availability of substitutes, and (4) oil imports in relation to domestic consumption.

The consumer entity modeled in this chapter is a generic importer of petroleum. We do not differentiate among oil-importing countries. The purpose is to obtain insights regarding the implications of worldwide exchanges of petroleum for the major trading entities, not for any individual country. The structure of the model is generic; the parameters can be altered to approximate individual countries.[2] The consumers can be disaggregated. But this much is indisputable: oil-importing countries ascribe some vulnerability to their demand for Gulf oil.

8.1 Modeling the Demand for Petroleum Imports

The consumer countries' petroleum imports from the Gulf region are modeled in four steps.

Demand
The first task is to estimate the total demand of the consumer countries. This is done by taking available information about projected consumer demand (under the assumption of 1970 prices), then modifying these projections and adjusting them with respect to the effects of prevailing oil prices and the impact of the availability of substitute energy. Substitute availability is also determined by the expected prices of petroleum. The higher the price expectation, and the lower the price of substitutes, the greater will be incentives for investment in alternatives to petroleum. The availability of substitutes will in turn reduce consumer oil demand. The result is a modified projection of consumer oil import demand. Figure 8.1 depicts these relationships.

The base demand series employed is a projection made by the OECD in a study of energy outlook for its member countries to 1985.[3] The OECD made three projections—an initial forecast predicated on about $3

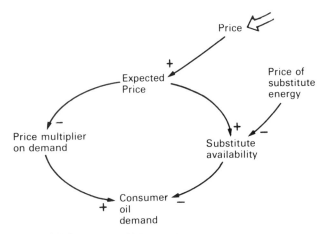

Figure 8.1 Consumer oil demand

per barrel and pre-1973 GNP growth rates and two other projections based on $6 and $9 in constant U.S. dollars.[4] Each projection assumes (1) the unimpeded functioning of a market environment, (2) no changes in the countries' energy policies except marginal adjustments to market conditions imbalances, and (3) the appropriateness of forecasts for GNP growth made prior to the price increases of 1973.

The initial OECD projection is used as base demand series. That projection relies in turn on earlier studies provided by member countries.[5] The GNP growth rate imbedded in the individual members' projections and adjusted by the OECD secretariat constitutes the core of the OECD's own initial demand forecast.[6] These growth rates are the following: for the United States a 4.3 percent increase is anticipated for 1971 to 1980 and 4.0 percent for 1980 to 1985. The rates are higher for Japan (8.1 percent for the first period and 7.4 percent for the second), the European Economic Community countries (5.2 percent and 4.8 percent) and other OECD European countries (5.1 percent and 5.2 percent).

In 1977 the OECD presented revised estimates of energy prospects based on a lower estimate of GNP growth.[7] The new forecasts result in lower demand and production estimates. Since the newer study employs higher oil prices than the pre-1973 prices of $3 per barrel, there is no direct comparison between the 1974 and 1977 revised forecasts of demand and supply for 1977 prices. For this reason we chose to employ the 1974 estimates rather than the downwardly revised 1977 forecasts. Our purpose is to initialize the entire model at as close to 1970 values as possible, with the effects of price increases generated endogenously. Demand and production forecasts that use higher prices would generate marked inconsistencies and biases in the IPE model.

The distinctive feature of the consumer demand formulation in the consumer sector of the IPE model is the adjustment of the base OECD projection by oil prices (determined endogenously to the model) and the price of alternative sources of energy (set exogenously). The modification is done by adjusting the base series according to additional updated information on oil prices and substitute prices in a simple multiplicative formulation.[8] The base consumer oil demand series is presented in figure 8.2. The final, adjusted demand series is endogenously determined for the IPE model as a whole.

The oil price multiplier on demand shows the influence of oil prices and price changes upon the demand for petroleum. This influence is based on two major factors: the ratio of the base price (set at the 1970 value of $1.80) to expected price (a smoothed value of price over a three-year period) and an exponent (analogous to a price elasticity) that repre-

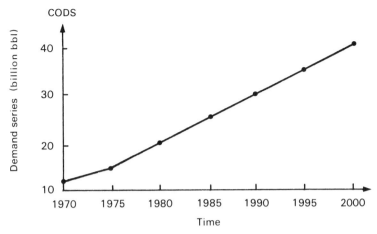

Figure 8.2 Consumer oil demand unadjusted series

sents the impact of that ratio set at 0.3. This impact constitutes our effort
to denote a moderate influence of change in price on demand.[9] In addition there is an adjustment period of five years over which the impact of
price on demand makes itself felt. The substitute availability multiplier
for demand reflects the impact of alternative sources of energy upon the
demand for petroleum.

The next step is to model oil production from domestic sources in consumer countries. The difference between demand and production will
then yield that amount of petroleum that must be imported if prevailing
demand is to be met.[10]

Production from Domestic Sources

Oil production from domestic sources in the consumer countries is subject
to the same constraints and processes described in chapter 4 with respect
to supply in the Gulf region. Domestic sources provide the consumer
countries with their own resources and the possibility of making a decision to expand these further. Such expansion, however, is constrained by
the actual volume of oil-in-place, production costs, and depletion. As
with demand an empirical data series for domestic production is used as
the base for modeling supply and modified by explicitly taking into account the effect of price on production and the available knowledge regarding depletion effects.[11]

The base production series (or domestic supply) series, also an initial
OECD projection, is set at 4.2 billion barrels per year in 1970, increasing
to 6.1 billion barrels by 1985.[12] The OECD projections take into account
the relationship between cost and price, specified according to individual
OECD regions. It is assumed that the price of domestic oil is allowed to

rise to the oil import prices. The production estimates for individual countries are drawn from separate studies.[13] The base production series are presented in figure 8.3.

The price and depletion adjustments made on the base production series together result in a final, revised estimate for consumer oil production. Depletion effects are included in the specification of consumer supply from indigenous sources in terms of the depletion associated with the fraction of reserves recovered at any point in time. The higher that fraction recovered, the greater will be the depletion effects, and the lower oil production will be, all other things being equal. This relationship is noted in figure 8.4. Reserves are initialized at 200 billion barrels in 1970. In the absence of empirical data on the influence of technological change, the effects of technology on domestic production are incorporated through the assumptions in the OECD supply forecasts and in the specified depletion effects.[14]

Figure 8.5 represents the two separate, but multiplicative, influences on consumer oil production: price and depletion effects.

Elasticities of Supply and Demand
Price multipliers on demand and on production represent the impact of expected price on consumer oil demand and domestic supply (produc-

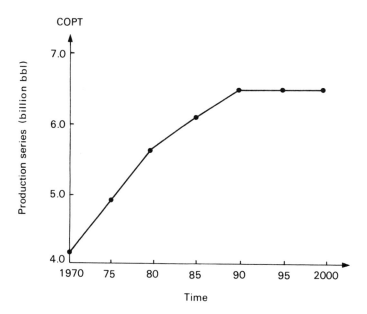

Figure 8.3 Unadjusted series of consumer oil production

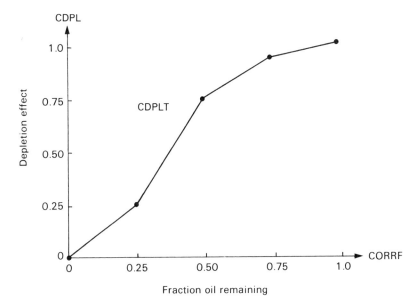

Figure 8.4 The depletion effect

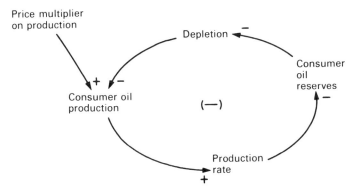

Figure 8.5 Consumer oil production

tion). These coefficients represent price influences from the levels assumed in the OECD forecasts and a way of incorporating price elasticities into the production and demand formulations. A price elasticity of demand (or supply) coincides with a percentage change in price. A higher expected price will simultaneously stimulate production and depress demand, and by extension higher domestic production and depressed demand will act to decrease the consumers' import demand. Conversely a lower price will ultimately increase consumer import demand through the same process. Expected price is the exponential average of actual prices over the previous three-year period.

The OECD study employs the elasticity coefficients provided by the National Petroleum Council in 1972.[15] The demand elasticities used in the OECD projections for 1980 and 1985 are differentiated by region and industrial sector. For the oil sector the OECD stipulates a long-run demand elasticity for the United States of 0.25 in 1980 and 0.3 in 1985. For Canada the elasticities are also 0.25 and 0.3, but for Europe they are 0.23 for 1980 and 1985, and for Japan 0.25 for each of these two years. Long-run elasticities imply price expectation increases sufficiently in advance for consumers to make the necessary adjustments.[16] Without providing supporting explanations, the OECD study states that domestic elasticities were not used directly.[17]

The long-term perspective for the oil-exporting countries of the Gulf region we used in the IPE model is set at 0.3 for both supply and demand elasticities, corresponding to the general thrust of the values in major oil models.[18] The supply coefficient is similar to that of Kennedy (1974), the midpoint in Kalymon (1975), close to, but slightly lower than, the Pindyck (1978) coefficient for the $6 per barrel case, and midpoint in the Ezzati (1976) coefficients for industrial and developing countries. Chapter 10 examines the implications of different elasticities. The coefficients used in the reference simulation are changed alternatively to higher and lower values and their affect on the behavior of the model identified accordingly.

Import Demand
Consumer oil import demand is simply the difference between demand and production from domestic sources. That difference yields total demand for imports which refers to a worldwide demand rather than to a demand from a particular region.[19] Actual data from the Gulf region are used to determine the fraction of total demand met from Gulf sources. It will be recalled that the oil producers in the Gulf area controlled 65.4

percent of total world exports of crude petroleum in 1978. Existing estimates suggest that the Gulf fraction of the world export market in 1980
and 1985 will be between 0.58 and 0.68, with the more likely range between 0.62 and 0.64. Such estimates are necessarily "soft." Production capacity for the Gulf for 1980 and 1985 was available only in the 1977 CIA
energy report, which indicated a capacity approximately equal to total
world imports predicted in the two OECD studies. Yet the CIA predicted
tight supplies by 1985 using different demand predictions. No trade
figures were given by the CIA, but, by adding estimated OPEC production and non-OPEC developing countries' supply to approximate world
imports, the 1980 Gulf export fractions can be determined (assuming
Saudi Arabia's production is 8 mb/d). For 1985 the range is 0.58 to 0.68 if
the extreme cases are assumed (demand and production matched, high-
low, low-high) or 0.64 to 0.62 if production is positively correlated with
demand (matched high-high, low-low).[20]

With this information, import demand is then divided between demand for imports from the Gulf and non-Gulf supply, with the implicit
assumption that consumers will not treat the Gulf as a residual supplier
but maintain stable market shares with their purchases. Computationally,
however, consumer non-Gulf supply is treated as a residual.

Assuming no changes in other variables, the higher the consumer countries' demand for oil, the greater will be the demand for imports. However, if the production from domestic sources increases, imports will
decline. Of course expected price always influences demand and supply.
Figure 8.6 depicts consumer oil import demand.

8.2 Strategic Vulnerability of Consumer Countries

At this juncture the consumer countries' major policy problems emerge:
an increased reliance on external sources for petroleum consumption involves greater dependence upon the suppliers. While the nature, extent,
and implications of such dependence have been widely debated, the general consensus in the consumer countries of the West is that excessive reliance on imports will result in greater susceptibility to external pressure, to
blackmail, or to the use of economic weapons for political purposes, in
addition to the political and economic consequences of an unintentional
supply interruption, as in Iran in 1979. The danger perceived is one of
strategic vulnerability. But no one has defined that vulnerability in a
comprehensive and precise way, nor is there a consensus as to its essential
features. Yet the perception of vulnerability is a real one, whether nations

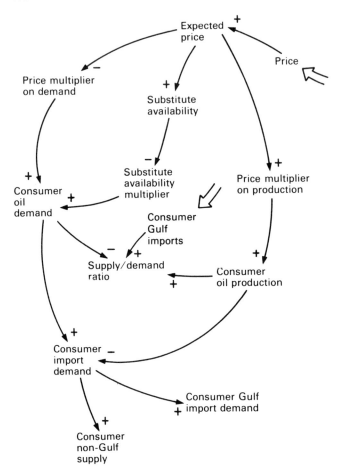

Figure 8.6 Consumer oil import demand

import by necessity or choice. Modeling basic features of such vulnerability is undertaken as part of the consumer sector to reflect the underlying political concerns emanating from increased imports of petroleum. In the absence of a theory of vulnerability, it is necessary to adopt a more pragmatic approach to concept specification.

Strategic vulnerability for consumer countries is modeled as a function of four variables, three of which are endogenous and the fourth based directly on empirical data: imports in relations to total consumption, availability of substitutes, essentiality of oil to industrial economies, and insecurity of available sources of supply.[21] All range between zero and one. The indicator—and its separate components—serves largely as a monitoring device to observe the effects of different oil import policies or of different energy policies.

Oil Imports
For some consumer countries the ratio of oil imports to total petroleum consumption is extremely high (Japan and Western Europe); for others it is considerably lower (the United States). But all consumer countries are concerned about the amount of their consumption met by imports. The broader political issue is to maintain unimpeded access to imports. In calculating the ratio, the consumer countries' domestic sources of production are taken into account. Thus any policy to increase output from domestic sources would obviously have a positive impact on this component of strategic vulnerability. Since in a crisis situation the OECD countries are formally committed to sharing the import shortfall equally, this aggregation of consumer countries is especially valid.

Substitute Availability
The availability of substitutes to petroleum is determined by an exogenously supplied price of alternative sources of energy (which at any given time essentially depends upon the current state of energy technology) and by the expected price of petroleum. If the expected price of petroleum is high, there will be a greater inclination to make investments in alternative sources of energy. Despite the rhetoric and the initial analyses there remains considerable uncertainty about the effects of oil prices on incentives for increasing investment in alternative sources of energy and the exact price of presently available alternatives for commercial use. There are even greater uncertainties about the appropriate analytical perspective in modeling the impact of oil prices on substitute availability for the consumer countries' strategic vulnerability. There is a necessary sim-

plification and a candid acknowledgment of uncertainty in both functional form and data base.

Figure 8.7 presents the series for the price of substitute energies which reflect an initially declining price of substitutes (due to an assumption of rising investment in alternative sources of energies and to technological development) followed by a leveling off (due to an assumption that major developments will occur by 1985).

Essentiality of Oil

Perhaps more fundamental than the availability of substitutes is the importance of petroleum to industrial economies. On this issue there is no controversy. The ratio of petroleum consumption to total energy consumption provides a clear and unmistakable indicator of essentiality. As alternatives to oil become increasingly available, that ratio will decline, marking a reduction in the role of petroleum. Essentiality is closely related to substitute availability. Both variables are included to allow for the differentiation between the extent of reliance on petroleum and the ease with which a switch to alternatives can be made.

The empirical base for the essentiality of oil is from three sources: (1) historical data for 1971 to 1976, (2) two OECD studies for forecasts to 1980 and 1985, and (3) projections for 1985 and 2000 by the Ford Energy

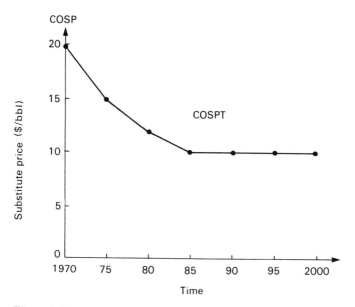

Figure 8.7 Price of substitute source of energy

Policy Project and by the Workshop on Alternative Energy Strategies. The evidence converges to 0.5 until 1985, indicating about half of the consumer countries' energy requirements will be met by petroleum. Toward the year 2000 that figure is projected to decline. But there is greater uncertainty surrounding the 2000 projections, for which the espoused assumptions are questionable. For this reason we adopt the more conservative estimates of essentiality of oil to consumer states and set that figure at 0.55 in 1970, declining to 0.45 by 2000. Clearly it will be simple to test the implications of different coefficients and table functions to demonstrate the extent to which international oil exchanges are affected by different assumptions about the essentiality of oil.

Insecurity of Supplies
The assessment of supply insecurity is basically a judgment, informed by intelligence reports and political analysis. However, an empirical indicator is necessary. Supply insecurity is set as the ratio of consumer Gulf oil imports to consumer oil demand over a four-year period.

Strategic Vulnerability Indicator
A change in any of the four variables determining strategic vulnerability could increase or decrease that vulnerability, depending on the direction of the change. The outcome indicator is computed as an average of the four values. In the absence of a theory of strategic vulnerability, developing and manipulating indicators is the best that can be done. Figure 8.8 presents the overall characteristics of the specifications for strategic vulnerability. The influence of current price and price expectation should be noted along with that of the price of substitute energies, in which the consumer governments may intervene to change their strategic vulnerability. Each of the determinant components are potentially manipulable, but to a different degree and at varying costs. For example, a decision could be made to limit imports from the Gulf, to increase investments in alternative sources of energy, or to reduce total energy consumption, and in each case there would be an impact upon the calculations of overall strategic vulnerability. Each decision involves government intervention in structural relationships.

The combined causal diagram for the entire consumer sector is presented in figure 8.9. The role of price, price expectation, and the price multipliers (elasticities) on demand and on production (as domestic supply) should be noted. The combined computer flow diagram for the consumer sector is given in figure 8.10.

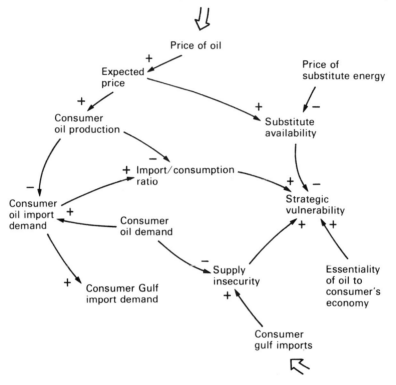

Figure 8.8 Strategic vulnerability

8.3 Linkages between the Consumer Sector and the IPE Model

Two variables are critical inputs into the specifications of the consumer sector. First is the price of oil. Price influences consumer oil demand as well as consumer oil production. The effects are mediated through the price multipliers. The second input is consumer Gulf oil imports, which represent actual imports from the Gulf in contrast to consumer Gulf oil import demand. Actual Gulf imports are a direct input into the strategic vulnerability calculations. The major contribution of the consumer sector is in deriving demand for Gulf oil and providing the framework for examining the strategic vulnerability that accompanies such demand.

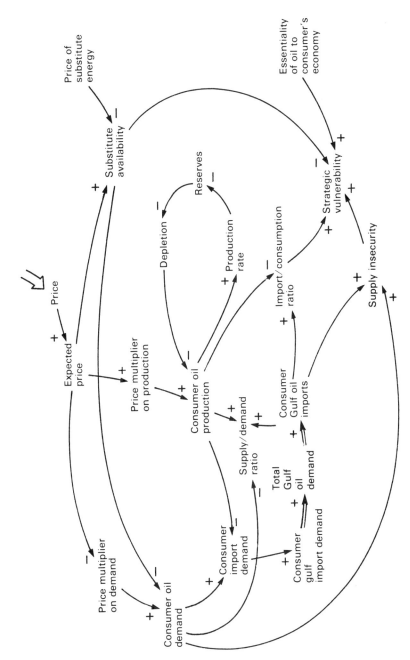

Figure 8.9 Causal loop diagram of the consumer sector

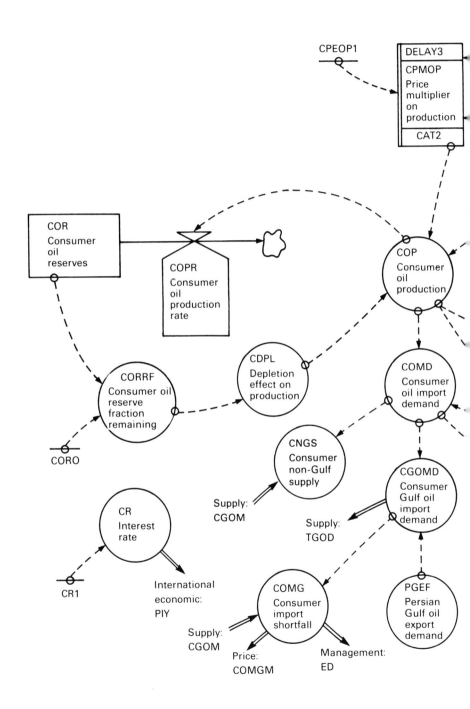

Figure 8.10 Computer flow diagram of the consumer sector

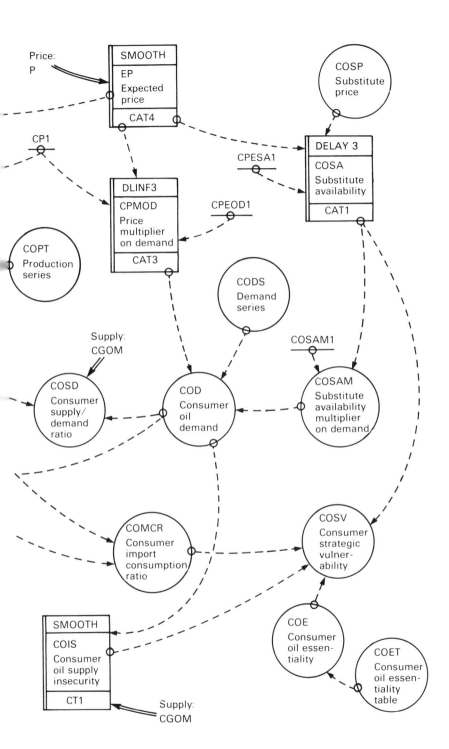

9
The International Economic Sector: Balance of Payments

The reciprocating effects of financial flows generated by the sale of petroleum on the balance of payments of the consumer countries and the foreign investments of the oil-exporting countries are calculated in this chapter. By seeking to model the recycling of petroleum revenue, major linkages are determined for the IPE model as a whole. The computations in this chapter follow conventional balance of payments accounting procedures.

9.1 Monitoring Financial Transactions

A country's balance of payments is an official record of all international transactions. It consists of two accounts: the current account and the capital account. The current account represents all trade in goods and services and other short-term transactions, including government exports and grants. The capital account includes private and government short- and long-term loans. The basic balance in international economic transactions is the sum of the current account and the capital account.

The current account is determined by the consumer countries' trade balance plus the difference between repatriated corporate profits and the producer countries' income from investments overseas not reinvested but repatriated. The balance of payments accounting for consumer states is not modeled in its entirety; only those features directly involving the exchange of goods and services with the oil-exporting countries of the Gulf region are included.

The capital account is the difference between the producer countries' investments in consumer economies and in their own countries, less the oil companies' direct oil investments from their working capital located in the oil-importing countries. Again these calculations are simply accounting relationships, but they yield a fairly comprehensive indication of international financial flows at any point in time.

A third balance of payments variable, the cumulative basic balance, is computed for monitoring purposes. It is set to zero during the first simulation period and then allowed to accumulate as financial inflows and outflows are recorded. This concept is unique to the IPE model and does not have an analogue in conventional balance of payments accounting. It consists of the cumulative deficit and the corresponding accumulation of foreign assets by creditor nations.

Figure 9.1 presents a diagram of the balance of payments relationships from the perspective of the consumer countries, mapping the financial interdependence between the oil-importing and the oil-exporting countries and the influence exerted by the oil companies. The higher the consumer countries' oil expenditures, the more negative the impact on their balance of payments position. For this reason the magnitude of imports and investments from the oil-exporting countries can potentially redress the balance. Similarly the impact on the consumers' balance of payments will be favorable to the extent that the international oil companies increase their repatriated profits and reduce their investments in the oil industry overseas.

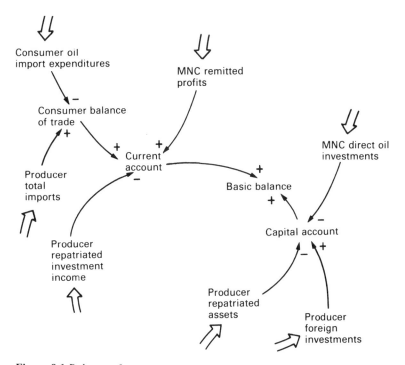

Figure 9.1 Balance of payments

Since in the accounting of international transactions a country's payments position must always be balanced, the important issue is whether existing flows can be maintained and whether trade exchanges indicate viability rather than marked imbalances. The interdependence generated by oil imports is corroborated by the fact that the consumers' balance of payments are influenced not only by their oil bill but by what the oil-exporting countries choose to adopt for foreign investments and trade policies and what the oil companies do to repatriate their profits.[1] Corporate decisions have been made in the Finance Sector as they bear directly upon the policies that underlie oil production, as depicted in the supply sector. The effects of the producer countries' foreign investments are now examined.

9.2 Foreign Investments of Producer Countries

From the oil-exporting countries' perspective the predicaments posed by extensive oil income (viewed as the surplus above current foreign exchange requirements for imports) has been regarded as potentially, or partially, resolved by recycling petroleum revenues. The oil producers' foreign investments are a direct input into the consumer countries' capital

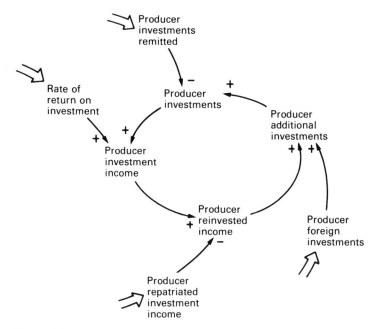

Figure 9.2 Producer countries' foreign investments

account. The producers' imports of goods and services are an input into the consumers' balance of trade, and any repatriated income from producer investments in the consumer country economies affects the consumers' balance of payments current balance.

Modeling the foreign investments of the producer countries involves a major positive feedback relationship, representing the initial producer foreign investments, the reinvestments in the economies of the consumer countries, and the additional investments. The oil-exporting countries' foreign investments generate reinvested income which in turn is combined with new foreign investments to create additional investments overseas, which, as the process continues, increases the overall foreign holdings of the oil-exporting countries. The size of remitted investments and the repatriation of investment income eventually serves to reduce the overall holdings overseas. As long as income is reinvested, there will be an increase in the producer countries' overall foreign investments. However, repatriation and remittances both are leakages from increasing producer investments abroad.[2] Figure 9.2 depicts these relationships.[3]

9.3 The Structure of International Economic Transactions

Figure 9.3 links the producer countries' investments overseas and the consumer countries' balance of payments situation. This is the heart of the recycling problem: the producer countries' foreign investments affect the consumers' capital account on the balance of payments, in that the higher the level of these investments, the more favorable it is from the perspective of the consumer countries. Certainly there are potential problems associated with the stability of these commitments. The producers' remitted investments also have a direct, but negative, impact in that they appear as a debit on the consumers' capital account. Similarly the repatriation of investment income on the part of the producer countries has an unfavorable effect on the current account of the consumer countries. Remitted corporate profits are a credit in the balance of payments of the home country through the current account calculation.

The most important determinant of this accounting from the consumer countries' perspective is the magnitude of their oil import expenditures. This alone generates a problem to which recycling has been posed as a potential solution. The requirement for such a solution is a robust interconnected structure of financial intermediaries, the major elements of which are characterized in figure 9.3.

A complete diagram of the equations for the international economic sector is presented in figure 9.4. This sector provides essentially an ac-

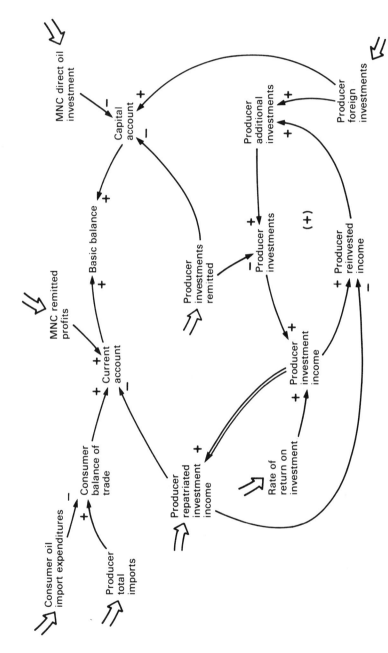

Figure 9.3 Causal loop diagram of the international economic sector. Double lines indicate intersector flows.

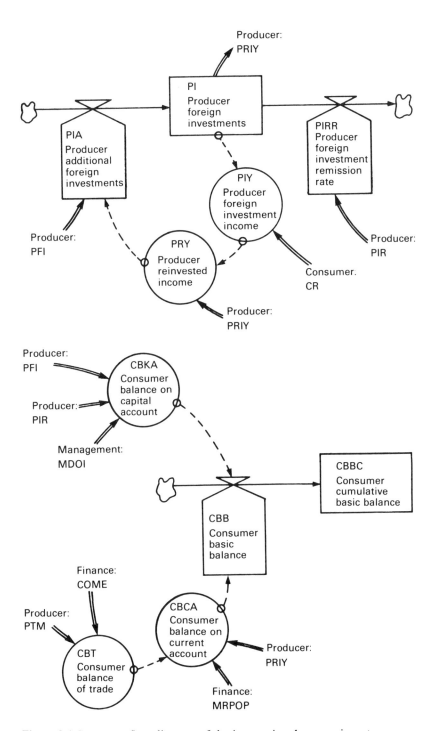

Figure 9.4 Computer flow diagram of the international economic sector

counting function that reflects behavioral relations determined in other sectors. There are no endogenously generated processes specified to this particular sector that require independent validation. The most important function of the international economic sector is to calculate the net as well as the gross flows of income between consumer and producer countries and register the effects of such flows upon one group's balance of payments and the other's investment patterns. But the processes generating the flows measured in this sector are specified in other sectors of the integrated model.

9.4 Linkages between the International Economic Sector and the IPE Model

Although the major relationships are between the oil-exporting and the oil-importing countries, the decisions of the oil companies concerning remission of profits and reinvestments in the oil industry also have direct bearing on the balance of payments of the consumer countries. For this reason corporate policies are also taken into account in the IPE model.

The IPE model is closed by tracing the producers' transactions as they affect the consumers' balance of payments, and by the same token the consumers' oil import expenditures as they affect the producers' economies and their investments overseas. The processes that underlie the major linkages for the IPE model are essentially generic processes. In both respects the international economic sector summarizes the transactions of the world petroleum exchange system.

III
The Results

10
The Results of the Reference Case

The IPE model adopts a systemwide perspective on the exchanges between producers and consumers of petroleum and their interactions with the oil companies. It represents not only the characteristics of the oil market but many features of global exchanges that are not developed in other models.[1] The IPE model provides a general framework for analyzing the flows of oil and of payments across national boundaries and identifying their worldwide repercussions. These flows generate a robust structure of global interdependence.

When fully integrated, the seven sectors of the model together present a general international perspective on oil issues. The reference simulation of the IPE model employs the historical (actual) tax rates from 1970 to 1978 as the major exogenous input into price determination and by extension to generating overall model behavior. The integrated model generates results not identical to those arising from a sector-by-sector analysis.

10.1 The Central Importance of the Producers' Tax Rate

Recall that price is set as a function of the tax rate of the oil-exporting countries, the markup of the international oil companies, and the costs of production. Markup incorporates some of the supply and demand relationships in the oil market and the influence that consumers exert upon the exchange, both in generating the demand for oil and oil imports and in making investments in domestic sources of oil and alternative sources of energy. Markup is not always a negligible component of price.

Different price policies—given different tax rates—generate different interdependencies throughout the structure of exchange. There will be different short-term and long-term effects and differences between final outcomes at the terminal year (2000) as well as along the paths to that point. Some features of the exchange are invariant and not subject to short- or long-term changes, and others are more variable. Different tax policies will affect all participants, but to different extents and at different times.

The basic reference simulations utilize a tax rate composed of known real values from 1970 to 1978 and projections of the 1978 tax rate at a constant level to the year 2000.[2] This reference rate is used only for experimental and comparative purposes. It can be regarded as a prediction only if one accepts some very conservative assumptions about the future: (1) that the Gulf producers will refrain from increasing taxes faster than the inflation rate, given the extent of the gains that had already been made by 1978, (2) that prevailing supply and demand relationships will not change dramatically relative to present trends, and (3) that the degree of cohesiveness among the suppliers in 1978 will not change.[3] The reference case is therefore a useful policy base against which to compare the effects of alternative tax rates and policies and different consumer response policies, as well as alternative supply and demand relationships.

10.2 Results of the Integrated Model for 1970 to 2000

Part II presented some partial, sector-specific links for key variables in the model. Each sector was examined as an individual component of the model. Even when two or more sectors were interfaced, the resulting simulation yielded only partially interconnected behaviors. The integrated results are generated by the model as a whole, in terms of a dynamic simulation structure with endogenous feedback relations for each major variable.

10.3 Demand

Recall that demand specification is in two parts: total demand for oil (COD) and demand for oil imports (COMD). Each responds to changes in oil prices and reflects the ways in which consumer countries adjust to the suppliers' taxes. The increases in oil prices in 1973, continuing to 1978, have had a strong impact on oil demand (COD). Consumer countries have attempted to adjust their overall demand, which increases by 5 billion barrels from 1978 to the terminal period 2000. Of course consumer countries can change the ways in which demand is met, for example, by increasing their non-Gulf share of consumption. Market conditions and prevailing prices may force non-Gulf suppliers to increase their production and meet more of consumer import demand.

The demand for imports (COMD) reflects some of these trends, but the patterns are more uniform and more consistent. Pre-1973 prices result in import growth. The subsequent increase in the tax rate stabilizes consumer imports, which persist until the early 1980s, but, since further growth in demand cannot be met from domestic sources, the trend cannot be sustained, and there is notable import growth beyond 1985. This result is due initially to increases in domestic production and subsequently to the constraints imposed by limited domestic reserves. In the reference case the consumer countries draw upon their reserves at a very high rate.

Over time the percentage of supply other than non-Gulf imports as a percentage of total demand (COSD) ranges between 75 percent in 1970, 74 percent in 1978, and 64 percent by 2000. This is the ratio of imports from the Gulf (CGOM) plus domestic production (COP) to total demand (COD) and can therefore be interpreted as the shortfall in the meeting of demand by consumers' production and the Gulf and a shift to a non-Gulf production. Consumer non-Gulf imports (CNGS) are specified to emulate producer production; as such, they increase in the 1970s, are stable in the early 1980s, and increase dramatically from 1985 to 2000, when they reach 7.5 billion barrels.

Recall that in the model the OECD projections (CODS) are adjusted by the price of oil and the availability of alternative energies to yield consumer demand (COD). The result is a substantially more conservative demand forecast. Figure 10.1 presents both the OECD demand series and

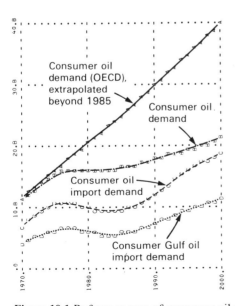

Figure 10.1 Reference case of consumer oil demand (billion barrels)

the modified endogenously determined forecast, as well as consumer import demand and import demand from the Gulf region.[4]

The differences between Gulf oil production (POP) and the consumption of producer states (PCD) is for export. Consumer imports from the Gulf (CGOM) more than double in magnitude between 1970 and the end of the period. In 1970 the consumers imported from the Gulf 4.17 billion barrels, by 1990 they will import 7.2 billion barrels, and by 2000 over 11.2 billion barrels, total imports being 6.9, 12.1, and 18.7 billion barrels, respectively. This expansion of imports clearly has critical economic consequences.

10.4 Supply

The supply specification in the integrated IPE model entails oil production in the exporting countries, production from domestic sources in the consumer countries, and supply from non-Gulf sources. Consumer oil production is also based on an OECD projection adjusted by the impact of oil prices as computed in the model (in contrast to the OECD price of $3 per barrel) and the impact of the depletion of domestic reserves. Again there are substantial differences between the base OECD projection and consumer oil production generated by the model. As figure 10.2 indicates, we forecast a substantial decline in production from domestic sources of

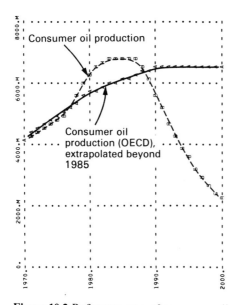

Figure 10.2 Reference case of consumer oil production (million barrels)

oil in the consumer countries, dropping dramatically after the mid 1980s, so that by the end of the period, in 2000, the levels of domestic production are close to half of those in 1970.

The differences in the two production projections arise in part due to the impact of oil price on production, induced supply effect, and depletion. A higher induced supply effect means lower reserves, and lower reserves mean greater depletion and lower final production.

The adjustment of the consumer countries to the tax rate of 1973 to 1978 entails attempting to cut demand, cut imports of oil, and to increase production from domestic sources. Production constraints in the consumer countries make it even more necessary for the oil-exporting countries to increase their production over the long run. The ratio of consumer oil import demand to the total oil demand (COMCR) reveals that the initial expansion of domestic production and curtailment of imports result in a decline in the ratio of import demand to total oil demand. But as constraints imposed by the size of domestic reserves become apparent, import demand increases and by the last decade of the period exceeds by far the level of imports during the earlier years. What is important here is the phasing and timing of adjustments in the Gulf region to the consumer countries' responses to price changes.

Oil production involves exploration and development. Each is affected by prevailing prices since that determines how much is bought and sold, the extent of investments in the industry, and the degree to which there are incentives for the development of alternative sources of oil or of energy. Table 10.1 summarizes the reference case projection of the overall trends in the oil discovery rate and the deposits or undiscovered oil (PUO) in the Gulf. The reference scenario makes considerable claims upon reserves (PRIOP) and causes a growing decline in undiscovered oil. Table 10.1 also indicates that the price increases of 1973 to 1978 result in a decline in oil production (POP), due to a marginal drop in production by 1980 related to changes in demand and consumer production. The discovery rate (PODR) grows toward the end of the 1980s as greater claims on Gulf production sources are made and actual production takes an upswing.

This process results in dramatic increase in capital and equipment (POC) from 1980 on. Oil capital increases respond to increased capacity utilization which accompanies increased production and steadily increasing costs. Effective investments in development contribute to this growth in capacity. Due to some decline in deposits (PRIOP) investments in exploration are made to generate greater recoverable oil-in-place and thus increase the rate of discovery (PODR) sharply by the end of the 1980s.

Table 10.1
Reference case of supply

Time	PUO[a]	PROIP[a]	POP[a]	PODR[b]	POC[c]
1970	500.00	350.00	4.283	1048.1	2.692
1971	498.98	346.02	4.746	977.5	2.640
1972	498.03	341.59	5.055	933.5	2.820
1973	497.10	336.80	5.510	930.1	3.261
1974	496.16	331.56	5.950	971.8	3.766
1975	495.15	325.93	6.363	1059.9	4.324
1976	494.03	319.99	6.613	1190.1	4.897
1977	492.77	313.97	6.664	1350.1	5.487
1978	491.35	308.05	6.555	1517.0	6.011
1979	489.76	302.38	6.338	1665.2	6.396
1980	488.04	297.02	6.085	1775.5	6.619
1981	486.23	291.91	5.929	1840.4	6.719
1982	484.38	286.96	5.832	1865.6	6.809
1983	482.51	282.08	5.801	1866.3	6.966
1984	480.65	277.20	5.833	1858.7	7.232
1985	478.79	272.26	5.918	1856.2	7.630
1986	476.93	267.20	6.081	1868.8	8.171
1987	475.05	261.98	6.272	1904.5	8.863
1988	473.11	256.56	6.606	1969.1	9.714
1989	471.10	250.74	7.074	2068.7	10.806
1990	468.97	244.56	7.486	2219.4	12.257
1991	466.66	238.01	7.976	2431.7	13.955
1992	464.11	231.09	8.474	2709.4	15.734
1993	461.25	223.83	8.935	3055.2	17.547
1994	458.01	216.32	9.366	3467.2	19.444
1995	454.34	208.60	9.781	3937.3	21.398
1996	450.17	200.72	10.184	4457.9	23.382
1997	445.46	192.71	10.575	5025.6	25.393
1998	440.16	184.62	10.956	5641.3	27.431
1999	434.23	176.46	11.329	6309.5	29.500
2000	427.59	168.31	11.627	7036.7	31.577

[a] In billion barrels.
[b] In million barrels.
[c] In billion 1979 dollars.

The historical tax rate effectively creates a process that places great strains on reserves and necessitates increases of investments in exploration. In the reference case there is an almost full utilization of oil capacity throughout the whole period.

The initial effect of the new tax rates in 1973 on desired additional production capacity is an immediate decline from the early 1970s to the end of the decade. Additions to capacity depend partly on desired oil production capacity (as distinct from the desired additional capacity), and the desired capacity is directly responsive to the ratio of expected demand to desired capacity utilization. Thus changes in demand have an immediate impact on desired capacity and by extension on the required additions to capacity. Furthermore the depreciation of existing production capacity necessitates in itself replacement of capacity.

Trends in desired additional production capacity (DPOPCA) and the loss of production capacity (POPCD) are presented in figure 10.3. The loss of capacity is a function of capital depreciation. Following an initial low level of desired additions to capacity, there is in the early 1970s a sharp increase, but the increase represents an immediate response, possibly an overreaction, that then exhibits a gradual downward trend throughout the 1980s decade. The last ten years of the period reveal a new increase in additional capacity, followed by stabilization. Despite consumer responses and producer adjustments, the volume of desired addi-

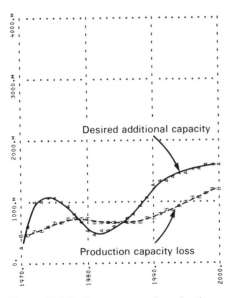

Figure 10.3 Reference case of production capacity (million barrels)

tions to production capacity by the end of the period is in keeping with the amounts required in the early 1970s prior to the initial declines. By contrast, the loss of capacity (POPCD) due to depreciation and costs per unit capacity remains fairly stable throughout the period. Higher capital costs toward the end of the period result in slight increases in capacity loss, but on the whole these factors do not appear to generate significant constraints on oil production or on utilization of capacity.

10.5 Price

Recall that the final price of oil is endogenously specified. Table 10.2 presents the price of oil per barrel (P), the tax rate (POTR), oil production costs (PMOC), and the companies' markup (MNCM). The tax rate is exogenous and historically determined from 1970 to 1978. All the other variables are responsive to changes in interactions among the major entities in the world oil market and to supply and demand relationships. Note the incremental changes in price due to markup and the changes in cost. Cost includes both discovery costs and production cost per unit capacity— which is small in the Gulf but cannot be ignored. Since the tax rate in the reference simulation is set at the 1978 level to the end of the period, the price of oil increases only slightly in comparison to the growth between 1973 and 1978.

In the reference case the companies' markup (MNCM) is small relative to the other components of price. Recall that markup represents the effects of four factors: the base markup in 1970 ($0.82 in 1970 dollars), the difference between desired and actual capacity utilization, the impact of the decline rate and the tightness of the market as indicated by the consumer oil import gap.[5] The slight markup decline following the initial price increases of 1973 is due in part to an initial consumer adjustment to oil prices by reducing their requirements (hence a lower consumer import gap demand and capacity utilization discrepancy which declines due to a decline in demand). But import cuts, which occur partly because of consumer adjustments to changing market conditions set during the 1970s, cannot be sustained. Domestic production falls as reserves are depleted, and demand increases due to economic growth. As these adjustments take place and a new equilibrium is reached, corporate markup stabilizes throughout the entire period.

In 1970 markup was $2.12 in 1979 terms. It was increased by about $.10 fifteen years later and then stabilized $.50 above the 1970 level for the remainder of the period. These changes appear minor, but they become important as the volume of trade grows. In a later chapter we see that,

Table 10.2
Reference case of price and components of price in 1979 dollars

Time	P	MNCM	PMOC	POTR
1970	5.018	2.1161	.5922	2.310
1971	5.543	2.4356	.6849	2.422
1972	6.700	2.5859	.7773	3.337
1973	6.302	2.5745	.8696	2.858
1974	14.440	2.5573	.9620	10.921
1975	17.096	2.5342	1.0546	13.507
1976	18.365	2.4841	1.1503	14.730
1977	17.992	2.3888	1.2466	14.357
1978	16.966	2.2813	1.3435	13.341
1979	16.924	2.1848	1.4410	13.298
1980	16.956	2.1184	1.5391	13.298
1981	17.049	2.1156	1.6348	13.298
1982	17.172	2.1432	1.7307	13.298
1983	17.311	2.1859	1.8267	13.298
1984	17.458	2.2369	1.9226	13.298
1985	17.602	2.2855	2.0185	13.298
1986	17.757	2.3413	2.1171	13.298
1987	17.897	2.3823	2.2158	13.298
1988	18.069	2.4561	2.3146	13.298
1989	18.256	2.5434	2.4138	13.298
1990	18.380	2.5684	2.5135	13.298
1991	18.483	2.5733	2.6112	13.298
1992	18.579	2.5705	2.7098	13.298
1993	18.674	2.5664	2.8096	13.298
1994	18.766	2.5562	2.9109	13.298
1995	18.858	2.5453	3.0138	13.298
1996	18.954	2.5368	3.1187	13.298
1997	19.055	2.5313	3.2256	13.298
1998	19.162	2.5285	3.3349	13.298
1999	19.273	2.5283	3.4466	13.298
2000	19.375	2.5160	3.5610	13.298

depending on the tax rate, markup can be as high as 50 percent of the total price. Markup as a percentage of price declines as oil prices increase—due to cuts in imports and reduction of capacity utilization. Indeed markup drops by $.35 between 1976 and the early 1980s when growth in imports contributes to a recovery and a slight increase in markup. By then relatively unconstrained capacity utilization and consumer import demand prevail, and the corporations' markup is forced up through the influence of the tight supplies in oil trade.

10.6 Consumer Import Payments, Corporate Profits, and Producer Oil Revenues

The consumer countries' expenditures on oil imports, the exporting countries' oil revenues, and the companies' oil-related profits mirror directly how much oil is bought and sold. The companies' investments in exploration and development reflect the adjustment to the interactions of buyers and sellers and the extent of its impact on oil production and utilization of capacity.

Consumer oil import expenditures (COME) depend upon the price of oil and the amount imported. Domestic sources of oil supply, the price of substitutes, and the extent of responsiveness to price change affect consumer countries' adjustment to changes in oil prices. This adjustment may take many forms, most notably a cut in imports, but there are constraints on the extent of such cuts. The jump in consumer oil expenditures apparent in 1974 (which in real terms is about $50 billion over the previous year) and the subsequent increase in the next few years are direct initial responses to the change in the tax rate. The present discounted value of the consumers' import bill in 2000, using a 5 percent rate of interest, is $50.3 billion in 1979 dollars.

The immediate consequence of a price increase is a growth in the revenue of the oil-exporting countries (POY). This represents a fourfold revenue increase in real terms (from $15 billion in 1973 to $63 billion the next year). The cumulated present discounted value of future income streams for the oil-exporting countries is $1,171 billion, again using a 5 percent interest rate in 1979 terms. Recall that income specification takes into account oil taxes and royalties, as well as costs of oil production. The actual taxes collected depend upon the rate per barrel and the volume of consumer imports of oil. The gains and losses of producers and consumers are both dependent on price and volume of trade.

Corporate profits (MOP) are the differences between the payments of

the consumer countries and the oil costs of the corporations plus the taxes
of the producer countries. The companies' profits more than tripled in the
1970s. Investments in exploration and development are made from corporate profits. The present discounted value of corporate oil profits is $6.5
billion for the year 2000.

Figure 10.4 shows the trends in consumer oil import expenditures, producer oil revenue, and corporate profits connected with petroleum trade,
and figure 10.5 gives their present discounted value. Oil payments and
producer revenue rose sharply from 1973 to 1978, followed by a slight
tapering off until the mid-1980s. This initial decline is due to adjustment
efforts by consumer countries to cut back on oil imports, but these efforts
cannot be sustained over the longer range due to the size of their own
reserves and the attendant constraints on expanding domestic production. The long-term effects of the current tax rate are an expansion of
consumer expenditures on oil imports and a growth in the producer countries' oil revenues.

For the international oil companies the trends in real profits also reflect
only marginally the increase in oil price of 1973 and the subsequent additions. The companies draw upon their profits to meet the costs of oil production and the expansion of oil facilities. Any increases in production or
exploration costs make themselves apparent by cutting into repatriated
profits. These factors create a leveling off of profits during the decade of
the 1980s. The subsequent increase in consumer oil imports—due in part
to limitations on domestic production—combined with the effects of earlier corporate investments in the oil industry generates an expansion of
production and production capacity and greater oil sales. As a result there
is a slight growth in corporate profits during the decade of the 1980s,
while leveling off during subsequent years.

In the reference case reinvested profits grow dramatically during the
1980s and 1990s, because of increases in consumer demand, oil imports,
and costs. Corporate investments are then allocated to development investments and exploration activity. Investments increase more steeply
than demand because of the influence of production cost per unit capacity, which increases. Exploration investments are a function of the
amount of oil remaining (deposits) and are related to discovery costs, actual production, and to the actual and desired decline rates. Both types of
investments also increase over time, but from different levels and at different rates. Both are affected by demand, but in different ways, and they
impact upon different aspects of the production process and supply relationships. Toward the end of the period the strain on existing oil supply

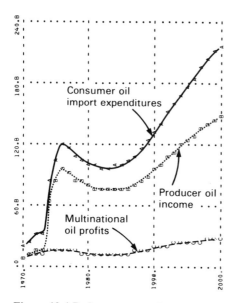

Figure 10.4 Reference case of consumer payments, corporate profits, and producer revenue (billion 1979 dollars)

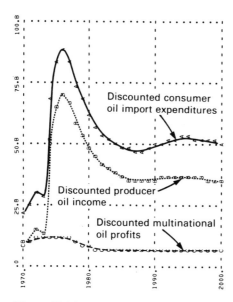

Figure 10.5 Reference case of present discounted value of future income streams (billion 1979 dollars)

from Gulf sources becomes apparent, and the corporations' concerns are reflected in an increase in desired discoveries. This increase is the outcome of exploration investments. There is a growth in exploration during the last decade that reflects the desire to expand reserves.

10.7 Producer Oil Revenues and Investments

Larger revenues from petroleum sales have direct effects on the producer countries and international financial transactions. Oil income, which is endogenously generated and depends upon the tax rate and the volume of oil trade, grows dramatically from 1974 to the end of the decade. The relative decline in the early 1980s, due to initial reductions in consumer oil imports, is reflected in the revenues of the oil-exporting countries. But because of the consumers' ability to sustain the curtailment of their imports, petroleum sales increase to the end of the period, with a corresponding growth in the income of the oil-exporting countries. In real terms revenues from oil sales will increase more than tenfold.

Trends in producer domestic investment demand (PKID) and imports of goods and services (PTM) are presented in figure 10.6. Of course they depend on oil revenue. The present discounted value of the producers' domestic investments in 2000 is $17.9 billion in 1979 dollars, and for their imports of goods and services it is $8.3 billion, again with a 5 percent interest rate.

Since domestic investments depend upon the initial capital stock and the country's capacity to absorb additional investments, there will not be marked annual changes reflecting the impact of oil revenue. The most notable effect is due to the price increases of 1973, and the rest of the decade reflects the new prices; the remainder of the period indicates the adjustments in the producer countries' economies. The growth rate in investment demand is greatest during the 1970s, followed by a more gradual increase throughout the following decades. On balance the producer countries' domestic investments will grow regardless of the tax rate at any point in time, but the levels and rates involved will differ. The same observation can be made with respect to the imports of goods and services. The immediate growth from 1973 to 1978 is followed by a generally upward, though more gradual, trend to 2000. The 1990s decade reveals an apparent adjustment reflected in a slower rate of growth.

Producer oil demand (PCD), which is responsive to population size and to domestic capital stock, an indication of industrialization, reveals a relatively lower rate of growth than the other development variables. The do-

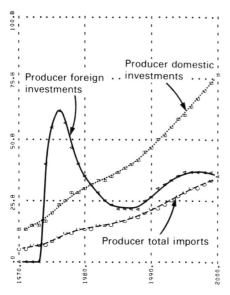

Figure 10.6 Reference case of producer domestic investments, imports, and foreign investments (billion 1979 dollars)

mestic consumption of oil will continue to be relatively small in any worldwide comparison—about 117 million barrels in 1970 to 399 million barrels by 2000—particularly when compared to other developments in producer countries. Because oil consumption in these countries does not depend immediately upon oil revenues, the mediating effects of industrialization and growth will in the long run have only indirect impacts. The more direct, sustainable, and dramatic effects of increases in oil revenue (in conjunction with population and industrialization) are seen in the demand for imports of goods and services and in domestic capital investments.

The producers' foreign investments (PFI) also grow dramatically. Income invested abroad is the difference between total oil income and expenditures on imports of services and on industrialization. While the trends in producer import demand are uniformly upward, those in foreign investments are more highly variable over time. The surge in the oil-rich countries' foreign investment from 1973 to 1978 reveals the initial constraint of absorptive capacity and relatively limited domestic investment demand. The decline in foreign investments during the 1980s reflects internal growth apparent in increases in imports of goods and services and in domestic capital investments and the decline in oil revenue.

10.8 Balance of Payments

The consumer countries' balance of payments takes into account economic transactions of the oil companies (as they make decisions regarding the repatriation of profits) and of the oil-exporting countries (as they make decisions regarding investments in the economies of the importing countries, or with respect to their imports of goods and services from abroad). Clearly the overall long-term balance of payments position of the consumer countries will deteriorate. But the tax rate increases in 1973 are only partly responsible. In real terms the basic balance in 1970 was in deficit at $6.9 billion and increased to $37.3 billion by 1978. By 2000 the basic balance will register a deficit of $128 billion. The present discounted value of the basic balance by 2000 is $29.8 billion. See table 10.3.

At any time the basic balance, the balance of trade, and the current account will be negatively affected—but for different reasons. In some years it will be because more oil is being imported, and in others, because less might be imported, but at a higher price. These trends are contrary to the commonly held belief that only the price hikes of 1973 create balance of payments problems for consumer countries. The economic predicament of importing countries transcends the events of 1973. It is a strong feature of consumer countries' dependence on petroleum imports. In chapter 12 we confirm this result.

The importing countries' balance of trade (CBT) is a more precise indicator of the economic interrelationship between oil payments and commodity exports to the producer countries. These exports directly offset the oil import bill, and the trade balance reveals additional information regarding the consumer countries' economic predicament. In 1970 the trade balance was $15.5 billion in deficit; by 1978 it worsened by about $81 billion to a total of $97 billion in 1979 dollars. This is due almost entirely to consumer oil import expenditures, since the imports of the producer countries were only $5.43 billion in 1970 and increased to $12 billion in 1978. By 2000 the producer countries would be importing about $36 billion worth in 1979 terms. The present discounted value of the balance of trade is $42 billion.

The important difference between the trends exhibited in the balance of payments and those in the trade balance lies not only in the overall patterns to 2000 but in the year-to-year fluctuations as well. The initial decline in the trade balance of the oil-importing countries is followed by a recovery due to the consumers' cuts in imports and increases in producers' own imports of goods and services. There is a brief period of im-

Table 10.3
Reference case of consumer balance of payments, in billion 1979 dollars

Time	CBBC	CBB	CBCA	CBKA	CBT
1970	.0	−6.92	−6.92	.000	−15.47
1971	−7.6	−8.65	−8.94	.000	−19.64
1972	−19.8	−14.49	−14.79	.000	−26.62
1973	−35.0	−14.61	−14.28	.000	−27.07
1974	−53.7	−22.88	−62.52	39.301	−76.19
1975	−78.8	−27.04	−82.67	55.205	−97.14
1976	−108.1	−31.06	−93.90	62.456	−108.61
1977	−141.2	−34.57	−91.93	57.060	−106.15
1978	−177.3	−37.26	−83.42	45.944	−96.76
1979	−215.5	−38.97	−79.79	40.671	−92.17
1980	−255.3	−40.39	−75.79	35.267	−87.32
1981	−296.5	−41.87	−73.11	31.073	−84.28
1982	−339.3	−43.57	−71.45	27.697	−82.46
1983	−383.9	−45.49	−70.79	25.095	−81.81
1984	−430.6	−47.67	−71.10	23.195	−82.25
1985	−479.5	−50.10	−72.19	21.843	−83.58
1986	−531.1	−52.87	−74.47	21.307	−86.26
1987	−585.7	−55.92	−77.25	21.014	−89.43
1988	−643.4	−59.45	−81.96	22.100	−94.93
1989	−705.3	−63.87	−89.32	24.994	−103.45
1990	−771.7	−68.53	−95.83	26.836	−110.54
1991	−842.9	−73.42	−103.40	29.467	−118.91
1992	−919.1	−78.60	−111.13	31.998	−127.45
1993	−1000.7	−84.03	−118.46	33.882	−135.40
1994	−1087.8	−89.62	−125.32	35.137	−142.68
1995	−1180.6	−95.38	−132.01	36.036	−149.71
1996	−1279.2	−101.38	−138.59	36.605	−156.55
1997	−1384.0	−107.63	−145.13	36.864	−163.24
1998	−1495.2	−114.17	−151.66	36.825	−169.80
1999	−1613.1	−121.03	−158.23	36.497	−176.26
2000	−1737.9	−127.93	−163.74	35.108	−181.32

provement favorable to the consumer countries during the early 1980s. However, the trends cannot be sustained over the long run.

The dramatic improvement of the consumers' capital account (CBKA) in 1973 to 1978 is due to the investments of the producer countries in the economies of the consumer countries. But the 1980s witness a slight decline, in real terms, due to the inability of the producer countries to sustain their investments abroad at the same high rate. Over time the producers' investments abroad do increase, contributing to the improvement of the consumers' capital account. This is due to surplus revenue in producer countries and the absence of any new investments by the oil companies over and above those derived from their petroleum-related profits. The present discounted value of the capital account, again with a 5 percent interest rate, is $8.1 billion.

The producer countries' domestic investment and imports also start to climb during the 1970s, as do additions to their investment income (PIA). But in this case the effect of consumer cutbacks in oil imports is relatively small—since this includes both their foreign investments and their reinvested income—and there is a continued, uninterrupted dramatic growth throughout the following decades. The patterns in the producer countries' reinvested income (PRY) are comparatively modest, but the trend is steadily upward. There is no interruption as a result of reductions in petroleum revenues, nor any sharp reversals or discontinuities. Some of this trend may be due to the interest rate in consumer countries (set at 5 percent per year), which compounds the accumulated investment income of the producer countries.

The growth in the producer countries' total investments (PI) is impressive. Since it is composed of these countries' investments in the economies of the consumer countries, in addition to their reinvested income it reflects the producer countries' overall investments. This variable, set at zero for 1970, indicates growth strictly as a function of petroleum trade. From $57.3 billion in 1975 producer investments increase dramatically to $254 billion in 1978, and more than triple that amount in the next ten years. By the end of the period producer investments are valued at $1,972 billion in 1979 terms. These figures are indeed overwhelming.

10.9 Strategic Vulnerability

Modeling the consumer countries' vulnerability associated with oil imports is one of the unique features of the IPE model. The historical tax rate, with the increases of 1973 to 1978, places the consumer countries in a position of greater vulnerability during the earlier years, but over time

there is a decline in vulnerability (see figure 10.7). This improvement is
due largely to greater availability of substitutes which offsets the more
immediate increases in the dependence on imported oil. Since the essentiality of oil declines to 2000 and insecurity of supplies also remains fairly
stable, despite fluctuations due to changes in consumer production, the
factor most responsible for the relative improvement in the consumer
countries' position is the increasing availability of substitutes. This is a
function of the expected price of oil and the price of substitutes. The tax
rate has an indirect, though definite, impact on the availability of substitutes. By the same token the most influential constraint on the improvement of the consumer countries' position lies in the ratio of import
demand to import demand plus domestic production. Since production
cannot expand significantly due to the size of the domestic reserves, this
ratio goes up over time, registering a definite, unfavorable impact on the
consumers' overall strategic vulnerability.

These changes in strategic vulnerability reveal that the constraints of
the earlier decades are replaced by solutions, such as lesser dependence
on imports from the Gulf, greater availability of substitutes, and a potentially lower import/consumption ratio. The relatively higher strategic vulnerability evident during the earlier period reflects extensive oil imports,
the absence of economically feasible substitutes, and continued dependence on Gulf sources. By the turn of the century, however, overall vul-

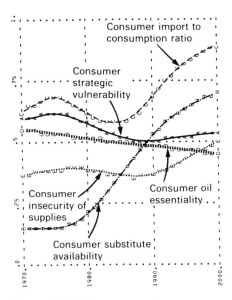

Figure 10.7 Reference case of consumer strategic vulnerability

nerability declines relative to the 1970s, leading to the attendant inference that long-term petroleum exchanges contribute to the stabilization of consumer countries' vulnerability at a level only marginally lower than in 1970. Indeed the most vulnerable period for the consumer countries is that immediately preceding the price increase of 1973. The early years of the 1970s represent a situation of greater vulnerability than at any time during the entire period. It is perhaps paradoxical that the increases in the tax rate have provided an improvement rather than a deterioration in the consumers' strategic position.

Early in the reference case the availability of substitutes is negligible, but as the price of substitutes drops over time, from $54 in 1970 to $27 by 1985 in 1979 dollars, alternative energies become increasingly available, with the greatest changes from 1980 to 1990. Insecurity of oil supplies, as the ratio of oil imports from the Gulf to total consumer demand, reflects the consumers' dependence on a politically insecure source. Thus the initial level of insecurity reflects the early high dependence of Gulf sources, and as that dependence declines—albeit marginally—insecurity decreases. The difference in these changes ought not to be overly emphasized given the metric involved. Changes later in the period reflect the decline in consumer oil production and concomitant rise in imports.

The most dramatic effect of the prevailing tax rates on the consumer countries pertains to the depletion of domestic resources. The depletion effect becomes particularly pronounced from the mid-1980s to the end of the period. This is because domestic reserves are being drawn upon at a rapid rate (due to increases in their own production), given the size of the reserves in the oil-importing countries. Offsetting production, however, is the expansion of imports by the mid-1980s. The trend continues to the end of the period. These factors reveal the type of constraints that will persist for the consumer countries.

10.10 Sensitivity Analysis

What are the effects of changes in some critical parameters?

If we allow the consumers to have greater reserves—initialized at 250 billion barrels in 1970 rather than 200 billion—and no other changes, we find that domestic production goes up accordingly, but the effects are so marginal that imports and import payments are not affected noticeably.

If we allow consumer countries to exhibit greater price responsiveness—by raising the demand elasticity to 0.5—and make no other changes, we find that there are significant effects. Oil demand is cut by 1.8 billion barrels in 1980. Gulf production is affected. Consumers import less, and their import bill is also smaller. The producers' revenues suffer accordingly.

The conjunction of larger consumer reserves and higher demand elasticity will result in a cut in demand, increased domestic production, reduced imports, and therefore a decline in Gulf oil production. The present discounted value of the producer countries' revenues will be cut by roughly one-fourth relative to the reference base. This means of course the import bill of the consumer countries will be improved accordingly.

If the consumer countries allow domestic production to be directly responsive to external prices, depletion possibilities, and known reserves of 200 billion barrels (the reference case), domestic production could increase late in the period, but by only 500 million barrels in 2000. Domestic production during the middle of the period would be substantially lower, decreased by as much as 2 billion barrels for several years, causing the present discounted value of the producers' revenue stream to increase by about 8 percent.[6] This amount is too marginal to have any notable effects on imports or on the production in the Gulf.

If consumer countries expand their reserves and allow production to be responsive only to external prices, then domestic production could increase by over 1.6 billion barrels in 2000, cutting into demand for imports and by extension into Gulf production. Production in the middle of the period would be higher but still well below the base case.

If the consumers expand their reserves, allow production to be responsive to price alone, and the demand elasticity is higher, then the effects become notable. Demand is cut by 4.5 billion barrels. Domestic production makes up for 1.6 billion barrels. The Gulf producers lose over 1.3 billion barrels, and substantial revenue losses occur as well.

What would happen if the non-Gulf suppliers expanded their production? The basic model stipulates that the non-Gulf suppliers meet a fixed percentage of the consumers' oil imports. If we allow the non-Gulf producers to exploit their reserves and respond to international prices and depletion possibilities, we find that the non-Gulf suppliers can play an important role. With initial reserves of 600 billion barrels and price elasticity of 0.3, non-Gulf supply is fairly stable. If non-Gulf sources were market-responsive, they would increase to 4 billion barrels by the end of the period rather than the 8 billion barrel level they reach when tied to Gulf production. (The effects vary by 1 billion barrels according to price elasticities of 0.1 and 0.5.) Since non-Gulf sources cannot expand significantly, the Gulf producers are always forced to take up the slack, so that their production reaches 15.5 billion barrels by 2000.

If all these changes occurred simultaneously—higher consumer reserves, greater demand elasticity, domestic production responding to prices, and the independence of non-Gulf suppliers—production in the Gulf would be lower by only 2 billion barrels in 2000.

Regarding the consumers' strategic vulnerability, the conclusion is this: None of these alternatives have strong effects on the consumers vulnerability position. Perhaps the only impacts pertain to the insecurity of supplies. Higher consumer reserves, greater demand elasticity, and the expansion of non-Gulf sources reduce the consumer countries' supply insecurity, since they are less reliant on Gulf imports. But overall the other components of vulnerability do not vary substantially.

This means that the basic reference case—as given in chapters 3 through 9—is sensitive to supply and demand parameter changes, largely as they affect domestic production and the expansion of non-Gulf sources. Parameter changes within the range tested here do not substantially influence the consumers' vulnerability position. This is because oil remains essential to industrial processes, and, under the tests made here, the incentives for expansion of alternative energies are not sufficiently strong to improve the consumers' overall position. The marginal reductions in imports during some years improve the consumers' position, but given the demand schedule and the nature of domestic reserves, these reductions cannot be sustained.

11
Evaluating the Results

This chapter examines the forecasts generated by the IPE model, the range of plausibility and some features of a possible future of the oil market.[1] The model results are compared, first, by checking wherever possible the simulated values with actual data (1970 to 1978), and second, by examining the reference case simulation with relation to the results of other oil models or related works. To a large extent the IPE model is based on empirical data. Nonetheless the results must be evaluated by assumptions, analytical structure, initial conditions, and endogenous behavior.

11.1 Simulation versus Actual Trends

Table 11.1 presents the real and simulated values for eight years of eleven critical variables in the integrated model which provide the basis for validation. The reference simulation tracks known (historical) values are indicated on an annual basis, from 1970 to 1978. All results are in 1979 dollars.

Looking at consumer countries' demand for oil, it is clear that the model output is very close to actual demand prior to the price increase of 1973 and off by more than 2 billion barrels subsequently. This difference, though not extensive, represents a deviation of 15 percent of total demand. It is difficult to determine how much of this difference is due to an initial overestimate of demand and how much is due to the underestimate of the consumers' response to the price increases in 1973. Recall that the model demand takes the OECD series and adjusts it by the effect of price and the availability of substitutes. Since the OECD estimates are in themselves price inelastic (see chapters 8 and 10), without these adjustments the demand estimates would have dramatically overshot actual demand. The OECD study also assumed a high rate of economic growth. With the recession of 1974 to 1975 the economy varied by 5 to 10 percent below the OECD estimate. This could account largely for the deviation in the model projection.

Consumer import demand (COMD) generated by the model is consistent with the actual values. The difference is less than 1 billion for six of the nine years. Differences can be explained by the year to year fluctuations due to inventory changes, price and demand expectations, and so forth. Imports are simply the difference between demand and domestic oil production. The model estimates production from domestic sources almost exactly from 1970 to 1972. The effects of the initial price increase result, by 1976, in about 1 billion barrels more produced (COP) by the model than by actual oil production in the consumer countries. This is due to the OECD projection of growing production even without price increases. If the model did not assume production would increase, all other things being equal, but instead allowed production to be affected only by price and depletion, the result would be the revised line for COP, which matches actual production data fairly closely.[2] Because both demand and domestic production are overestimated by 1977, somewhat offsetting each other, the result is about a billion barrel overestimate in the import demand of consumer countries.

Oil production in the Gulf is almost precisely estimated prior to the price increases. From 1975 to 1978 the model produces about 0.5 to 1 billion barrels more than do the Gulf fields. Again, although this amount is not large, it is extensive, given the total volume of oil production. Since demand is overestimated by the model, it probably is the source of the overestimate in Gulf production. In addition price controls on consumer countries' domestic production, notably the United States, increase demand, thus imports, and depress domestic supply.

The model also underestimates the oil consumption of the exporting countries. This can be explained by the fact that the coefficients in this equation are based on data for three specific countries, thus possibly generating an underestimate for the Gulf region as a whole. Further these coefficients were based on observations to 1973, which may cause underestimates of the effect of higher petroleum revenues in the later years.

The total investment of the producer countries (PI) is generated entirely by the model, in the sense that it derives from oil revenue and is endogenously computed. This includes their foreign investments as well as the returns on their investments overseas. There is an obvious underestimate (possibly a misspecification) in comparison with actual data prior to 1974. This may also be due to disparities within the Gulf, which are obscured by aggregation. Subsequently the model overestimates pro-

Table 11.1
Comparisons of model results and actual data

	1970	1971	1972	1973
Consumer oil demand (billion barrels)				
Actual	11.32	11.88	12.77	13.68
Reference	11.14	12.01	12.88	13.69
Consumer oil import demand (billion barrels)				
Actual	7.53	8.09	8.49	9.91
Reference	6.94	7.71	8.46	9.18
Consumer oil production (billion barrels)				
Actual	4.29	4.23	4.27	4.16
Reference	4.20	4.30	4.41	4.52
Revised[a]	4.20	4.17	4.14	4.11
Producer oil production (billion barrels)[b]				
Actual	4.37	4.44	4.80	5.70
Reference	4.28	4.75	5.06	5.51
Producer oil demand (billion barrels)				
Actual	0.111	0.121	0.136	0.160
Reference	0.117	0.121	0.126	0.131
Producer oil income (billions of 1979 dollars)[b]				
Actual[c]	16.83	19.79	19.73	27.92
Reference	9.55	11.01	16.21	15.12
Producer investments (billions of 1979 dollars)				
Actual	7.78	9.56	12.11	15.69
Reference	0	0	0	0
Revised[d]	0	0	0	0
Producer total imports (billions of 1979 dollars)				
Actual	9.18	10.12	11.76	14.43
Reference	5.44	5.74	6.12	6.53
Revised[d]	9.31	9.69	10.81	11.99
Consumer oil import expenditures (billions of 1979 dollars)[b]				
Actual	25.71	29.61	31.66	40.92
Reference	20.91	25.38	32.75	33.60

1974	1975	1976	1977	1978
12.99	12.56	13.41	13.69	14.00
14.47	15.19	15.75	16.09	16.23
9.63	8.61	9.64	10.04	9.73
9.84	10.41	10.78	10.85	10.66
3.99	3.88	3.90	4.12	4.41
4.64	4.77	4.97	5.24	5.57
4.10	4.11	4.15	4.25	4.39
5.74	5.18	5.86	5.95	5.47
5.95	6.36	6.61	6.66	6.56
0.181	0.217	0.266	0.294	0.362
0.136	0.141	0.147	0.154	0.161
79.98	70.58	85.58	80.95	80.44
62.89	83.36	94.85	93.31	85.30
40.03	57.53	74.76	80.68	na
7.0	57.3	118.3	186.6	253.7
4.56	42.57	81.95	120.70	152.08
19.76	29.99	39.92	48.05[a]	na
7.14	8.55	9.85	11.04	12.00
14.96	25.49	34.18	41.05	45.20
95.75	82.73	97.07	100.95	88.24
33.33	105.69	118.47	117.18	108.76

Table 11.1 (continued)

	1970	1971	1972	1973
Consumer balance of trade (billions of 1979 dollars)[b]				
Actual	− 16.53	− 19.49	− 19.89	− 25.17
Reference	− 15.47	− 19.64	− 26.62	− 27.07
Revised[d]	− 11.60	− 15.69	− 21.94	− 21.61

Sources: *BP Statistical Review of the World Oil Industry,* 1972–1978; *International Financial Statistics,* International Monetary Fund, May 1979; *International Petroleum Encyclopedia,* 1977.

[a] See text.

[b] Variable labels refer to model variables as defined in chapters 3 to 9. Actual values include adjustments due to differences in definitions.

[c] Producer petroleum exports.

[d] Revised. See text.

ducer investments. Historical values are markedly lower in 1976 and 1977 than in the model. One important explanation is that the model does not incorporate the constraints of absorptive capacity. The model's estimates of producers' domestic investments do not reflect the limitations on the ability of the producer countries to carry out investment programs. The figures generated indicate what is available for investment after other commitments are taken into account, but there is no specification that they are effectively invested. The revised line represents an adjustment of the equation specifying producer imports of goods and services, because the reference case dramatically underestimates producer import demand, a higher PYMRI (the coefficient for oil income effect on imports) was tested. As can be seen, the results are better, though still high. This may be due to a high estimate of the rate of return on investments, which is set at 5 percent.

The model significantly underestimates the imports demand of the producer countries—by about $4 billion in the early 1970s and $12 billion by 1974. This can be accounted for largely by the fact that the equations do not include a large-scale importation of weapon systems, sophisticated technology, or other goods that have been bought by Iran and Saudi Arabia in recent years. These countries' import bill for noncommercial goods has been extensive and is not generated by our model—nor is it intended to be. Only an increase of the income coefficient in the import demand equation by a factor of ten produced the close fit seen in the revised estimate. This result confirms that the producers changed their behavior substantially after the 1973 price increases.

A much closer correspondence between simulated and actual values is apparent in the consumer countries' payments for oil imports. However,

1974	1975	1976	1977	1978
− 76.00	− 52.74	− 57.15	− 52.90	na
− 76.19	− 97.14	− 108.61	− 106.15	− 96.76
− 68.37	− 80.21	− 84.28	− 76.13	− 63.55

the model underestimates the consumers' expenditures for several years until 1975. In general the model seems to underestimate the import payments of the consumer countries, while overestimating oil demand.[3]

There is a close correspondence in the consumers' balance of trade (CBT). There is an underestimate of $1 billion in 1970, becoming an overestimate of nearly $7 billion for 1972, and from 1975 a continuous overestimate. Since the balance of trade is the difference between the producers' imports of goods and services from the consumer countries and the consumers' oil payments, it reflects the trends in both variables, to some extent offsetting errors. Although better, the revised data still overestimate the later years.

In summary the sources of external error include OECD's economic growth assumptions, the optimistic OECD consumer production forecasts, and regression underestimates of the effect of oil income increases. (Note also that the model adjusts the tax rate only every six months, which causes some slight variations from historical data.)

11.2 Comparison with Other Models

In comparing the IPE model results with those generated by other models, several facts must be kept in mind: (1) the IPE model is a general model, while other oil models are explicitly partial equilibrium structures, (2) the model assumes a tax rate per barrel, but price is generated endogenously, (3) none of the existing models includes all the variables of the IPE model, so the comparisons are shaped by the limitations in the scope of other models, and (4) the IPE model does not disaggregate individual consumer or producer countries, thus precluding specific comparisons with models that address themselves to individual countries. The following comparisons are nonetheless comprehensive given existing evidence.

Table 11.2 compares the oil demand estimates of ten other sources, with the model's forecasts and actual data provided by British Petroleum (BP). Note that for 1975 the model demand is higher by 3 billion barrels than the historical data, although it is within the existing range of other

Table 11.2
Comparisons of consumer oil demand (COD) of Western Europe, Japan, and the United States (unless otherwise specified in billion barrels)

	1970	1975	1980	1985	2000
Reference analysis	11.14	15.19	16.15	16.35	21.03
Historical data[a]		11.32	12.56		
American Petroleum Institute	10.95	14.84	19.27		
Gay (1976)			Min[b] Max[c]		
		12.67	14.42 16.06		
Schurr and Homan (1971)[d]			14.48		
Kennedy (1974)					
Base case: $5.25 duty (PG)			12.74–12.92[e]		
$7.00 duty (PG)			11.83–11.97		
$8.75 duty (PG)			11.02–11.28		
OECD (1974)					
Base case: $3/bbl			20.0	24.7	
$6/bbl case			16.8	19.3	
$9/bbl case			14.9	16.9	
OECD (1977)					
Reference case			15.0	17.2	
Accelerated policy				14.6	
High growth				18.5	
Low growth				16.2	
CIA (1977)			14.3–15.3	16.8–19.2	
ITC (1977)			14.3	16.4	
WAES (1977)[f]					
Case C, C-1[g]				15.81	21.0
Case D, D-1[h]				15.51	17.41
FEA (1974)					
$3/bbl case				25.3	
$6/bbl case				20.1	
$9/bbl case				16.8	

Sources: American Petroleum Institute (1969); Central Intelligence Agency (1977); Federal Energy Agency (1974); Gay, as cited in Bhattacharya (1977); Kennedy (1974); Gersic and Deyman (1977); Organization for Economic Co-Operation and Development (1974 and 1977); Schurr and Homan (1971); Workshop for Alternative Energy Strategies (1977).
[a] *BP Statistical Review of the World Oil Industry* (1975).
[b] 2.8 percent growth in demand for oil, 1975 to 1980.
[c] 5.2 percent growth in demand for oil, 1975 to 1980.
[d] North Sea production not considered.
[e] Using base and high supply elasticities (0.33 and 0.67).
[f] Using all of North America instead of just the United States.
[g] $11.50 oil from 1977 to 1985; $17.25 oil from 1985 to 2000.
[h] $11.50 oil.

estimates. If we consider that the recession of 1975 to 1976 in the West caused a lower demand for oil, an event that the model does not incorporate due to its use of the OECD demand series, then the forecasts generated are consistent with empirical realities. By 1980 the model's estimate of 16.15 billion barrels is well within the middle of the range provided by other estimates. By 1985 the IPE is less than 10 percent below the projections that assume a relatively high price for oil and almost equal to the high figure for WAES in 2000. We must conclude that the model forecast of COD is on the high side. The difference in estimates is due to a variety of factors, including price and the effects of substitute availability.

In calculating the import requirements of consumer countries, we have taken account of production from domestic sources (COP). In 1975 the model estimate for COP is about 1 billion barrels more than the actual (historical) figure. By 1980 it is well within the range of other estimates. By 2000 the IPE model generates a forecast for consumer oil production lower by about 2 billion barrels than the WAES estimate. Thus in this respect our assessment of domestic production for the long range is somewhat less optimistic than the WAES study whereas it is well within the range of the estimates for 1980 and 1985 (see table 11.3).

Since consumer oil imports represent the difference between demand and production from domestic sources, it remains to compare the IPE forecast for oil import demand (COMD) with those generated by other studies. In 1975 the model's estimate was higher than the actual figure by 1.75 billion barrels. Subsequently the import forecasts of the model appear to be on the lower side of the range of estimates provided by other studies. It is between the CIA and WAES estimates for 1985 and higher by about 2 billion barrels in comparison with WAES's maximum for 2000. The differences are due largely to differences in oil demand estimates. In comparing these results, note the different price assumptions in the estimates of consumer oil imports (see table 11.4).

In terms of consumer imports from the Gulf region (CGOMD) the only explicit estimate is given in *Middle Eastern Oil and the Western World* (1971), which puts forth 3.85 billion barrels in 1975 and 4.86 billion barrels for 1980. The IPE model's forecast is 6.2 billion barrels for 1975 and 5.9 billion barrels for 1980. These figures are higher by about 2 and 1 billion barrels and reinforce upward estimates. The IPE model is distinctive in its projection of CGOMD since it generates price specific estimates.

Undoubtedly the raison d'être for these estimates is in part to determine the magnitude of consumer oil import expenditures (COME). In the IPE model these expenditures are simply the number of barrels times the

Table 11.3
Comparison of consumer oil production (COP) (billion barrels)

	1970	1975	1980	1985	2000
Reference analysis	4.20	4.77	6.30	6.85	2.31
Historical data[a]	4.29	3.88			
Gay (1976)					
Minimum			5.48		
Maximum			6.58		
Schurr and Homan (1971)[b]			4.46		
Kennedy (1974)					
Gulf duty:					
$5.25			5.84–6.68[c]		
$7.00			6.06–7.26		
$8.75			6.31–7.30		
Ezzati (1975)					
$10 OPEC take/bbl			6.86		
$12 OPEC take/bbl			7.13		
CIA (1977)			5.0	5.1–5.8	
ITC (1977)			5.67	6.57	
OECD (1974)					
Base: $3/bbl			5.7	6.1	
$6/bbl case			6.4	7.6	
$9/bbl case			7.0	10.6	
Ben-Shahar (1976)					
$3/bbl			3.0	3.0	
$4/bbl			4.0	4.0	
$5/bbl			6.0	8.75	
$6/bbl			6.2	10.25	
$7/bbl			6.3	11.25	
$8/bbl			6.5	12.25	
FEA (1974)					
$3/bbl				4.9	
$6/bbl				6.2	
$9/bbl				8.0	
OECD (1977)					
Reference case			5.3	5.9	
Accelerated policy				6.8	
High growth				5.9	
Low growth				5.9	
WAES (1977)[d]					
Case C, C-1[e]				6.8	4.6
Case D, D-1[f]				6.0	4.0

Sources: Ben-Shahar (1976); Ezzati (1975); Central Intelligence Agency (1977); Federal Energy Agency (1974); Gay, as cited in Bhattacharya; Kennedy (1974); Gersic and Deyman (1977); Organization for Economic Co-Operation and Development (1974 and 1977); Schurr and Homan (1971); Workshop for Alternative Energy Strategies (1977).

Notes to table 11.3 continued:

[a] *BP Statistical Review of the World Oil Industry*, 1975 (London: British Petroleum Co., 1976).
[b] North Sea production not considered.
[c] Using base and high supply elasticities (0.33 and 0.67).
[d] Using all of North America instead of just the United States.
[e] $11.50 oil from 1977 to 1985; $17.25 oil from 1985 to 2000.
[f] $11.50 oil.

price per barrel. The eventual oil bill of the consumer countries by 1980 is somewhat higher than the OECD (1977) estimate for its reference case, and by 1985, due to adjustments on the consumer side, the model forecasts an import bill close to most estimates for that year. It is almost equal to the OECD's own reference case estimate. On the whole therefore from the consumer countries' perspective our estimate of their expenditures on oil imports is relatively optimistic. These comparisons are presented in table 11.5.

For greater detail alternative estimates of the import/consumption ratio (COMCR) are presented in table 11.6. It is clear that our estimates for 1980 are among the lowest. The same is true for 1985. But by 2000 our projection is higher than the only explicit forecast, that of WAES. On balance, however, the IPE generates more conservative estimates in the shorter range than in the other models. This is due in large part to the more conservative (or optimistic) estimates of consumer oil demand. The IPE model reveals prospective adjustments due to price changes and the responsiveness of substitutes to price.

Petroleum Production and Supply
There exist only fragmentary longitudinal estimates of forecasts for oil production in the Gulf. This is due presumably to the belief that these fields are fully explored and their potentials known and/or to the fact that since 1973 the predominant preoccupation among oil production forecasters was with non-Gulf sources, those deemed to be more secure and accessible to the West than the Gulf. In addition it is considered more difficult to forecast Gulf production due to the role of politics in determining production levels. Several spot year estimates are presented in table 11.7, and the only existing trends over time are given in figure 11.1. In the short run we estimate a decline in Gulf production in the reference case in comparison with the other two forecasts. This finding results from several factors, including the availability of substitutes and of course the persistence of the 1978 tax rate to 2000, which is for experimental purposes only.

Table 11.4
Comparison of consumer oil import demand (COMD) (billion barrels)

	1970	1975	1980	1985	2000
Reference analysis	6.94	10.41	9.85	9.50	18.71
Historical data[a]	7.53	8.61			
Schurr and Homan (1971)[b]		8.01	10.49		
Abolfathi (1977)		8.60	10.10	12.00	
Gay (1976)[c]					
Maximum			10.58		
Minimum			7.84		
Kennedy (1974)					
Gulf duty					
$5.75			6.24–6.90[d]		
$7.00			4.71–5.77		
$8.75			3.36–4.71		
Murakami (1975)				7.15	
OECD (1974)					
Base case: $3/bbl			14.4	18.6	
$6/bbl case			10.4	11.7	
$9/bbl case			7.8	7.7	
OECD (1977)					
Reference case			10.4	12.1	
Accelerated policy				8.4	
High growth				13.5	
Low growth				11.1	
CIA (1977)			9.3–10.3	11.3–14.1	
ITC (1977)			8.7	9.8	
FEA (1974)					
$4/bbl				20.5	
$7/bbl				13.9	
$11/bbl				8.8	
Pindyck (1978)[e]				13.87–14.6	
WAES (1977)[f]					
Case C, C-1[g]				9.01	16.40
Case D, D-1[h]				9.51	13.41

Sources: Central Intelligence Agency (1977); Federal Energy Agency (1974); Gay, as cited in Bhattacharya (1977); Kennedy (1974); Gersic and Deyman (1977); Organization for Economic Co-Operation and Development (1974 and 1977); Schurr and Homan (1971); Workshop for Alternative Energy Strategies (1977); Abolfathi et al. (1977); Murakami (1975); Pindyck (1978).

[a] *BP Statistical Review of the World Oil Industry,* 1975 (London: British Petroleum Company, 1976).

[b] North Sea production not considered.

[c] Minimum using 2.8 percent growth in demand for oil, 1975 to 1980; maximum using 5.2 percent growth in demand, 1975 to 1980.

[d] Using base and high supply elasticities (0.33 and 0.67).

[e] OECD imports from OPEC.

Notes to table 11.4 continued:

f Using all of North America instead of just the United States.
g $11.50 oil from 1977 to 1985; $17.25 oil from 1985 to 2000.
h $11.50 oil.

In 1975 our estimate of producer oil production (POP) is 6.36 billion barrels, slightly lower than Murakami's assessment. By 1980 it is also about that amount, considerably lower than the CIA estimates. By 2000 it is 11.6 billion barrels, substantially higher than Pindyck's forecast for that year. This is also true of comparisons with spot estimates of production capacity.

Producer Oil Revenue, Foreign Investments, and Domestic Development
The variations in forecasts of producer countries' revenue accrued from the sale of petroleum are indeed profound. Since different initial prices are employed, as well as different assumptions regarding the price determination mechanisms, future oil prices take different paths. In some models an optimal price is identified. In other models future paths are delineated with no specific objective function that is maximized; rather price is determined by some market or other control mechanism.[4] In particular the income paths of Ben-Shahar's (1976) $8 per barrel and $11 per barrel cases straddle that of the reference case. The income path produced by Abolfathi et al. (1977) is about 20 to 40 percent higher, although that analysis includes sales to non-OECD consumers, while the IPE model does not. The results of Pindyck's model (1978) also bear similarities to that of the IPE model.

To compare more specifically the results generated by the IPE model with those of Pindyck (1978), table 11.8 presents the present discounted values for producer oil revenue in these two cases. Since Pindyck projects revenues for all the OPEC countries, and the IPE model refers only to the Gulf producers, some adjustment in the two series must be made to achieve comparability. Thus the last column of table 11.8 presents an estimate of the approximate Gulf fraction of the oil revenues projected by Pindyck for all the OPEC countries. Note the higher convergence between the two series during the terminal years.

There has been considerable speculation regarding the impact of higher oil revenues upon the producer countries' imports of goods and services. But there is only one forecast other than that generated by the IPE model. Abolfathi (1977) estimates producer imports of goods and services at $119 billion for 1980 and $181 billion for 1985, compared to $13.3 billion and $16.6 billion for the IPE model.

Table 11.5

Comparisons of total consumer oil import expenditures (billion 1979 dollars)[a]

	1970	1975	1976	1980	1985	2000
Historical data	43.71	140.48	162.53			
Reference analysis	34.8	178.0	197.9	167.0	167.2	362.6
Kennedy (1974)[b]						
Gulf duty						
$5.75				87.5–72.4		
$7.00				79.7–99.9		
$8.75				61.7–95.0		
OECD (1974)						
Base case: $3/bbl				100.1	129.5	
$6/bbl case				144.6	162.9	
$9/bbl case				162.9	196.3	
OECD (1977)						
Reference case				167.1	194.6	
Accelerated policy					135.1	
High growth					217.2	
Low growth					178.5	
FEA (1974)						
$ 4/bbl					168.5	
$ 7/bbl					199.2	
$11/bbl					198.7	
WAES (1977)[c]						
Case C, C-1[d]					201.1	549.1
Case D, D-1[e]					212.4	299.4

Sources: Bhattacharya (1977); Federal Energy Agency (1974); Kennedy (1974); Organization for Economic Co-Operation and Development (1974 and 1977); Workshop for Alternative Energy Strategies.

[a] With the exception of the historical and reference results, Bhattacharya (1977) and Kennedy (1974), all data has been derived by multiplying price times oil consumption. Thus the first three are closer to consumer expenditures, and the last four are more representative of producer revenues (not including delivery costs and a difference of about $1 to $2/bbl). This represents total imports, not just Gulf, and is not comparable to table 11.1.

[b] Using base and high supply elasticities (0.33 and 0.67).

[c] Using all of North America instead of just the United States.

[d] $11.50 oil from 1977 to 1985; $17.25 oil from 1985 to 2000.

[e] $11.50 oil.

Table 11.6
Comparisons for consumer import consumption ratio (COMCR) (percentages)

	1980	1985	2000
Reference analysis	61.0	58.1	89.0
Schurr and Homan (1971)[a]	72.4		
Gay (1976)			
Maximum	65.9		
Minimum	54.4		
Kennedy (1974)[b]			
Gulf duty			
$5.25	54.2–48.3		
$7.00	48.8–39.3		
$8.75	42.7–29.8		
OECD (1974)			
Base case: $3/bbl	71.6	75.3	
$6/bbl case	61.9	60.6	
$9/bbl case	52.7	42.1	
OECD (1977)			
Reference case	66.2	67.2	
Accelerated policy		55.3	
High growth		69.6	
Low growth		65.3	
CIA (1977)	65.0–67.3	68.9–70.9	
ITC (1977)	60.5	59.9	
FEA (1974)			
$3 case		80.7	
$6 case		69.2	
$9 case		52.4	
WAES (1977)[c]			
Case C, C-1[d]		60.0	78.1
Case D, D-1[e]		61.3	77.0

Sources: Central Intelligence Agency (1977); Federal Energy Agency (1974); Gay, as cited in Bhattacharya (1977); Kennedy (1974); Gersic and Deyman (1977); Organization for Economic Co-Operation and Development (1974 and 1977); Schurr and Homan (1971); Workshop for Alternative Energy Strategies (1977).
[a] North Sea production not considered.
[b] Using base and high supply elasticities (0.33 and 0.67).
[c] Using all of North America instead of just the United States.
[d] $11.50 oil from 1977 to 1985; $17.25 oil from 1985 to 2000.
[e] $11.50 oil.

Table 11.7
Comparisons for producer oil production (POP) (billion barrels)

	1970	1975	1980	1985	1990
Historical data	4.37	5.18			
Reference analysis	4.28	6.36	6.08	5.92	7.49
ITC (1976)			3.5–13.1	9.9–16.7	
Murakami[a] (1975)	4.05	6.5	7.92	9.34	
Pindyck[b] (1978a)			12.0–12.67	15.70–16.06	
CIA[c] (1977)				17.05–18.69	
Gay[d] (1976)			18.43–20.62		
Kennedy (1975)[e]					
Gulf Duty					
$5.25			5.44–4.78		
$7.00			4.05–2.81		
$8.75			2.70–1.10		
CIA[f] (1977)			12.1–12.3	12.8–13.7	
Pindyck[g] (1978a)					15.70–16.43

Sources: Abolfathi et al. (1977); Central Intelligence Agency (1977); Gay, as cited in Bhattacharya (1977); Gersic and Deyman (1977); Kennedy (1974); Murakami (1975); Pindyck (1978 and 1978a).
[a] Iran, Kuwait, and Saudi Arabia.
[b] OPEC required production.
[c] OPEC required production.
[d] Total OPEC supply.
[e] Using base and high supply elasticities (0.33 and 0.67).
[f] Persian Gulf production capacity.
[g] OPEC production capacity.

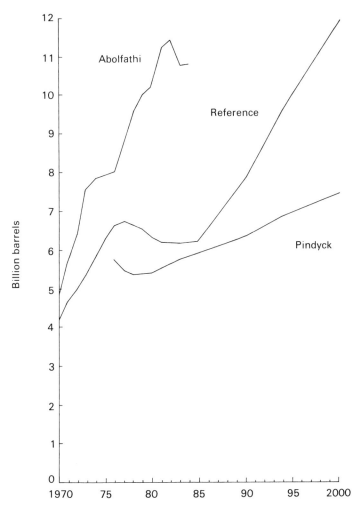

Figure 11.1 Producer oil production (POP)

Table 11.8
Pindyck optimal producer oil income versus the IPE model

	IPE model (1979 dollars)[a] Reference	Pindyck	Pindyck (1979 dollars)[b]	Approximate Gulf fraction of revenue[c]
1975	82.8	109.1	177.1	124.1
1976	90.4	103.6	131.4	92.1
1977	85.4	94.5	110.5	77.3
1978	75.2	83.7	100.4	70.3
1979	69.9	72.5	95.0	66.5
1980	64.6	61.9	92.1	64.4
1985	50.2	52.0	84.8	58.5
1990	48.2	55.5	78.9	55.2
1995	48.2	59.0	72.4	50.6
2000	44.6	56.2	66.0	46.3

Source: Pindyck (1978).
[a] Discounted at 5 percent, with 1975 = year 0.
[b] Pindyck uses two alternative discount rates, 5 percent and 10 percent; we have used 5 percent and tested with 10 percent and 20 percent. The general trends are similar. Results are available to the reader upon request.
[c] Gulf production divided by OPEC production (as shown in *BP Statistical Review of the World Oil Industry*, 1976 and 1977).

Existing results do not converge in terms of overall trends in forecasts of producer investments (PI). The Ben-Shahar (1976) analysis generates higher dollars available for total investment by the producer countries.

Only Ben-Shahar produces a time series forecast for the producer countries' foreign investments (PFI), which remains considerably higher than the IPE model's base forecast. Table 11.9 presents this forecast. Table 11.10 lists a set of spot estimates by Mikdashi of total producer investments (PI) for 1980 and 1985 under alternative assumptions.

Finally, it remains for us to compare our estimates for the producer countries' own consumption of oil (PCD) and those provided by other sources. Available estimates are in table 11.11. Clearly the IPE model consistently generates estimates that are lower than other sources following the price increases of 1973. For the early 1970s the model's estimates for producer oil consumption are consistent with the estimates of Abolfathi. But the latter generates a forecast that attributes greater oil consumption to the producers than in the IPE model.

Balance of Payments
Other than the IPE projections there currently exist very few long-term forecasts of the balance of trade for the oil-consuming nations, or of the

Table 11.9
Comparisons of producer foreign investments (PFI) (billion 1979 dollars)

	1973	1974	1975	1976	1977	1978	1979	1980	1985
Reference	0	39.3	55.2	62.5	57.1	45.9	40.7	35.3	21.8
Ben-Shahar[a]	7.86	77.0	75.4	74.0	69.2	70.8	64.3	54.9	48.7
Producer imports (PTM)									
Reference	6.5	7.1	8.5	9.8	11.0	12.0	12.7	13.3	16.6
Revised[b]	11.9	14.9	25.3	34.0	41.0	45.4	46.7	47.4	49.7
Ben-Shahar[a]	14.2	40.9	47.2	51.9	58.1	59.7	69.1	77.1	70.7

Source: Ben-Shahar (1976).
[a] Iran, Iraq, Libya, Saudi Arabia, and the "Persian Gulf emirates."
[b] See table 11.1.

Table 11.10
Comparisons of producer investment (PI) (billion 1979 dollars)

	1973	1974	1975	1976	1977	1978	1979	1980	1985
Reference	0	7.0	57.3	118.3	136.3	253.7	311.2	366.2	625.6
Ben-Shahar[a]	15.8	23.6	100.7	176.0	249.6	319.5	389.4	453.3	690.9
Mikdashi scenario[b]									
1								514.2	705.3
2								468.4	508.8
3								417.3	317.7
4								296.1	− 13.2
5								352.7	344.6

Sources: *Petroleum Economist* (February 1976), Ben-Shahar (1976).
[a] OPEC total international foreign reserve.
[b] Mikdashi scenarios (OPEC total international foreign reserve):

Consumer GNP growth	Income elasticity	Consumer energy supply	Non-OPEC imports	Oil export price basis
1 High	High	Low	Low	$10.46
2 High	Low	Low	Low	$10.46
3 Low	Low	Low	Low	$10.46
4 Low	Low	High	High	$10.46
5 High	High	Low	Low	$ 8.00

Table 11.11
Comparisons for producer oil demand (PCD) (billion barrels)

	Abolfathi	API[a]	Schurr and Homan	Reference
1970	0.111	0.261		0.117
1971	0.122			0.121
1972	0.135			0.126
1973	0.168			0.131
1974	0.209			0.136
1975	0.256	0.347	0.401	0.141
1976	0.304			0.147
1977	0.354			0.154
1978	0.407			0.161
1979	0.459			0.169
1980	0.515	0.497	0.630	0.176
1981	0.574			0.184
1982	0.648			0.192
1983	0.721			0.200
1984	0.789			0.209

Sources: Abolfathi et al. (1977); American Petroleum Institute (1969); Schurr and
Homan (1971).
[a] The Gulf excluding Qatar and United Arab emirates.

cumulative effects of their balance of payments position (Bhattacharya
1977, Rybczynski 1976). The existing spot estimates indicate some rough
correspondences for 1974, with a more dramatic deterioration forecasted
by the IPE model for the following year. These spot checks can hardly
constitute a basis for validation. The absence of balance of payment estimates for consumer countries makes it difficult to anchor our results in a
broader analysis.

11.3 Summary

This chapter compared the major results of the IPE model with empirical
data for 1970 to 1978 to indicate the extent of convergence with actual
trends. In general we found good fits for consumer import demand and
for consumer imports. Producer development variables were underesti-
mated, as was the size of the consumers' oil import bill. We then compared our results to find where in the spectrum of available information
these might be placed. From the perspective of consumer countries our
results are less pessimistic than those of other studies. The reference analysis projects a considerably lower future demand than by another model

and considerably more optimistic than the OECD studies. For producer countries our results are much more detailed than currently available elsewhere, thereby precluding any comparisons. Since the basic parameters of the producer equations are based on empirical estimates (pooling cross-sectional and empirical data), the key parameters are solidly grounded in the scene and the time frame utilized. The resulting interconnections with the remainder of the model generate behaviors that are extremely plausible.

IV
The Future

12
Comparing Alternative Policies: Politics and Economics

We now ask: If the oil-exporting countries change their tax rate—thereby inducing a change in price—what will be the effects on supply and demand relationships and the overall exchanges among consumers, producers, and international oil companies? Clearly there are numerous, alternative possible tax rates other than the six options in this book. These options reflect fundamentally different policy orientations. We emphasize the differences among the six types of tax rates. The final prices include the influence of corporate markup and production costs.

12.1 Alternative Futures

This chapter compares the results of the six policies directly and reveals their advantages and disadvantages for the participants in the world oil market. Table 12.1 summarizes the concerns of the producers, consumers, and oil companies. Also included are global concerns pertaining to the depletion of resources. However, whose resources are being depleted undoubtedly becomes a political issue. Table 12.2 presents the tax rate for the six policies in 1970 and 1979 dollars.

Conservative Tax Policy
By present standards a policy of doubling the 1978 tax rate in 1985 is conservative. The corresponding model-determined price would jump from $5.02 in 1970 to $30.78 in 1985. Although this tax rate is stipulated to remain constant beyond 1985, price would increase to $32.55 by 2000, since it responds to supply and demand relationships and increasing costs. In this case the producers obviously benefit the most relative to other policies, with the exception of the radical case. The producers receive the greatest amount of revenue for the least amount of oil produced, and their economies make the greatest gains. This is reflected in the value

Table 12.1
Select criteria for evaluating alternative oil prices

Consumer countries (importers)	Producer countries (exporters)	International oil companies (managers)
Demand for oil		Markup
Oil import expenditures	Oil revenue	Profits
Balance of payments	Imports of goods and services	Oil-related investments
Strategic vulnerability	Foreign investments	Oil-related costs
Domestic oil reserves	Domestic investments	
Energy substitutes	Oil reserves	

Note: These are not exhaustive criteria. Other variables in the model can be used for evaluation purposes.

of capital stock, the volume of imports, and their investments both domestically and abroad. The effects of the cut in consumer oil demand are more than offset by the increase in revenue per barrel.

But the impact on demand is important. The tax increase occurs just at the point where consumer reserves are strained and demand for imports grows again. This will create pressures on the exporters to increase their production. The tax increase serves to moderate this chain of events, due to its effects on consumer demand.

Conversely this scenario is poor for the oil companies, who are faced with a significant decline in demand and in oil sales midway through the period. Changes in petroleum trade have direct impacts on their oil profits. Corporate investments decline in response to the drop in demand and the new constraints on profits. But this response is of insufficient proportions to compensate for the reduction in profits.

A doubling of the tax rate has negative effects for the consumer countries, but there are also some long-term benefits. Despite lower oil trade, higher prices create a negative impact on the balance of trade. However, higher prices mean greater foreign investments, which cause the basic balance to be almost as favorable to the consumer nations as the reference case, and more favorable in the final years of the century. A high tax rate also means increased consumer oil production and therefore a lower strategic vulnerability. It is paradoxical that the higher the tax rate, the lower will be the consumers' strategic vulnerability in the longer run. So, too, there will be a greater availability of substitutes that will help to offset the constraints on domestic production.

The IPE model does not measure the more immediate and direct effects of higher oil prices on the economies of the consumer countries, but it is safe to suggest that a doubling of the tax rate would produce certain negative consequences in the short run, although some of the longer-term effects cannot be regarded as uniformly unfavorable for the consumers. In this tax rate scenario clearly the producers gain the most and the consumers and multinational oil companies benefit least.

Weak Tax Policy with OPEC Breakup in 1985

A cut in the tax rate in 1985 shows decided gains for the consumers and multinationals at the expense of the producer countries. In real terms this policy involves a final price change from $5.02 in 1970 to $16.97 in 1978 and down to $10.92 in 1985. The cut in the tax rate does not stimulate oil demand until late in the period, so that oil revenues are substantially smaller than in the conservative, radical, optimal, or reference cases. As a result the producers' real economic gains are significantly smaller. This is apparent in the developmental variables as well as in revenue and foreign investments. Oil production is among the highest relative to the other scenarios, indicating the dual losses to the producers—in claims on resources due to high production, and in revenue foregone due to lower taxes.

Similarly the oil company profits do not show much increase overall, although by the end of the period the amount of profits being repatriated is the largest among the scenarios examined because of the lag in oil investments. The revenue lost to the producers is the consumers' gain, as their import expenditures drop dramatically. In addition the favorability of the consumers' basic balance is second only to the no-OPEC scenario. The only real cost to the consumers is in terms of a greater strategic vulnerability, since lower prices mean more imports, lesser availability of substitutes, and higher imports relative to consumption. The lower the price of oil, the greater the vulnerability of the consumer countries.

There is little question but that the consumers gain from a halving of the tax rate in 1985, and the producers lose. The corporations' gain is small but growing at the end of the period.

No-OPEC Low Tax Policy

The no-OPEC case means a real decline in the tax rate. The real price of oil under these circumstances would drop from 1970 to 1978 and then increase gradually to $7.65 in 2000, due to the influence of supply and demand relationships and rising production costs. Over the thirty-year period this would represent a change of only $2.63 in constant dollars. What will be pushing up price then is the cost per barrel in conjunction with the markup of the oil companies, assuming minimal price responsiveness to demand.

Table 12.2
Tax rates for alternative policies (in 1970 and 1979 dollars)[a]

	1970	1971	1972	1973	1974	1975
Historical rate						
1970 Dollars	0.858	0.90	1.24	1.06	4.09	5.02
1979 Dollars	2.31	2.42	3.34	2.85	11.01	3.51
Conservative						
1970	0.858	0.90	1.24	1.06	4.09	5.02
1979	2.31	2.42	3.34	2.45	11.01	13.51
Breakup						
1970	0.858	0.90	1.24	1.06	4.09	5.02
1979	2.31	2.42	3.34	2.85	11.01	13.51
No-OPEC						
1970	0.858	0.81	0.74	0.62	0.50	0.45
1979	2.31	2.18	1.99	1.67	1.35	1.21
Radical						
1970 Dollars	0.858	0.90	1.24	1.06	4.09	6.62
1979 Dollars	2.31	2.42	3.34	2.85	11.01	17.82
Optimal rate[b]						
1970	0.858	0.90	1.24	1.06	4.09	6.62
1979	2.31	2.42	3.34	2.85	11.01	17.82

[a] Current dollar values are converted according to the *international monetary fund export unit value index for industrial countries.*
[b] Adapted from Pindyck (1978).

A no-OPEC situation puts the greatest economic burden on the producers, to the great benefit of both consumers and the multinationals. It indicates what would have happened had there been no collusion among the producers throughout this entire period. These results can also be viewed as an explanation, or a justification, for the producer countries' dissatisfaction with the price structure prevailing in the early 1970s.

Oil production is much higher than in other cases, the producers' revenues are dramatically smaller, and their foreign investments deteriorate to the extent that their oil revenues cannot even meet their own capital needs. The companies reap their greatest profits compared with other tax rate cases. However, there are soaring exploration investments late in the period, due to greater production in response to growing demand, which cause a significant decline in their repatriated profits. The consumers' economic benefits are clear. Their import expenditures are the lowest of the six tax rate situations, and their balance of trade is the most favorable. As might be expected, however, the consumers' strategic vulnerability is the highest in this case. Because their demand is unconstrained, they import more than in other situations, and their imports relative to consump-

1976	1977	1978	1980	1985	1990	1995	2000
5.48	5.33	4.94	4.94	4.94			
14.75	14.35	13.30	13.30	13.30			
5.48	5.33	4.94	4.94	9.88			
14.75	14.35	13.30	13.30	26.60			
5.48	5.33	4.94	4.94	2.47			
14.75	14.35	13.30	13.30	6.65			
0.45	0.41	0.36	0.36	0.36			
1.10	0.97	0.97	0.97	0.97			
			9.93	13.24	16.90	21.57	27.53
			26.73	35.64	45.49	58.07	74.11
			4.85	5.31	5.86	6.41	6.95
			13.06	14.29	15.78	17.26	18.71

tion rise. Substitute availability is retarded due to lower prices, and the uncertainty of supplies is greater because of the volume of imports from the Gulf. Consequently, while on strictly economic grounds consumers benefit most from this scenario, in a broader context it is clear that they incur great costs, particularly vulnerability.

Optimal Price

This price path stipulates that it is optimal for the producer countries to increase prices sharply and then reduce them before allowing a more gradual increase by 1985, continuing, also gradually, to the end of the period. In this way they make large immediate gains and forestall adjustments in demand, since demand is relatively inelastic in the short term. This policy alternative proceeds on the basis of the Pindyck (1978) analysis, determining the optimal price for OPEC in 1975 and the future price stream, given a 5 percent (alternatively 10 percent) discount rate in a monopolistic market.[1]

A tax rate that generates an optimal price in terms of the producer discounted value of future income stream yields results roughly similar to the reference case. The producers' revenues are high, and their economies benefit with less oil produced than in the reference case but more than in the radical or conservative scenarios.

This tax policy generates the lowest profits for the corporations, with the exception of the radical and conservative cases. The companies therefore are confronted with a very unfavorable situation. What might be optimal for the producers is among the worst possible outcomes for the corporations. The optimal price result casts doubt on the view that the companies and the producers' interests are identical.

Consumer oil import expenditures are higher by $110 billion in terms of the present discounted value of the entire stream over time, and the amount imported is smaller by about 2 billion barrels in the terminal year, relative to the reference case.

In this tax rate situation the producers again gain more—though not as much as in the radical case—relative to consumer countries and multinational oil companies. Given the time path of the tax rate, which climbs, drops, and then rises slowly, it seems likely that both consumers and corporations would be especially dissatisfied with a situation that deprives them of price stability.

Radical Tax Policy

A radical tax policy is predicated on the assumption that the producer countries will maintain an optimal price from 1970 to 1975 and then increase their tax rate by 7.2 percent annually to 1985 and by 5 percent annually to 2000.[2] It represents a fourfold increase compared with the reference case, threefold in comparison with the Pindyck optimal model. Likewise it is clearly much greater than the conservative policy. The price soars from $5.02 in 1970 to $30.26 in 1980, $50.18 in 1990, and a peak of $79.50 by 2000.

By all counts the producer countries gain most by a radical tax policy. This gain is not only in oil revenue, and attendant development, but also in the value of remaining resources. The greatest benefits are generated by the lowest production rates. Thus the producers obtain high oil revenues and conserve their resources.

The consumer countries incur dramatic costs, but there are benefits as well. Their capital account consistently registers a strong and forceful position, their strategic vulnerability registers a record low in comparison with other tax policies, and their dependence on external sources of petroleum is lowest. Thus in the long run the consumers obtain important gains from a radical tax policy. This is a controversial result since on economic grounds alone low prices are always better for consumers than higher prices. But low prices generate dependence on imports. This dependence creates greater strategic vulnerability. Lower prices mean that the producers do not have the resources to invest in consumer countries.

These investments are important to the consumers since, whatever the price of oil, the consumers import oil extensively and incur deficits on their current account. These deficits can be offset by the investments of the producer countries which appear as credits on the consumers' capital account.

The companies' oil profits are the lowest of any scenario due to lower demand, but the size of the profits remains large since the tax rate is passed on to the consumers in the price. In the long run, as imports level off, the radical tax rate gives the lowest profits of all the tax rate scenarios.

12.2 Market Adjustments

Figure 12.1 presents the trends in oil production for each of the five tax policies and the reference case. The no-OPEC case places the greatest claims on the production of oil since demand is relatively unconstrained by price, and production expands to meet that demand. Higher demand places pressures on supply that induce production expansion and increases in oil capital. Oil production in the no-OPEC case is much greater than in any other scenario throughout the 1980s and 1990s. By 2000 it is nearly twice the magnitude projected for the reference case. Such high

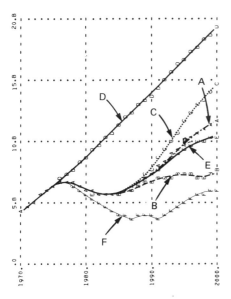

Figure 12.1 Producer oil production (billion barrels). A = reference case, B = conservative policy, C = breakup 1985, D = no-OPEC, E = optimal pricing, F = radical policy

production necessarily entails a much higher discovery rate. Furthermore by 2000 the decline rate is almost three times faster than in the reference case. The implications of these trends for oil reserves and deposits are dramatic. Recoverable oil-in-place declines sharply throughout the 1990s, and there is a substantial decrease in the pool of undiscovered oil. The decline rate, which is a function of the ratio of oil production to recoverable oil-in-place, exhibits a marked increase over time relative to the reference case.

The conservative tax policy creates stable production levels over time. This is because the consumers' immediate response to the 1985 price increase is to reduce demand and cut imports. Higher taxes will result in lower production relative to the other tax policies. As the volume of trade drops sharply, production capacity declines, as does the amount of capacity actually utilized. The political problem posed for the producers in this case entails a decision as to who will undertake production cutbacks and to what extent to coordinate production schedules. This situation necessitates continued cohesion among the Gulf producers and the implementation of disincentives for individual price cuts or unilateral production increases.

An optimal oil tax policy generates a production path not substantially different from the reference case because of the similarity of consumer demand in both cases. From a political perspective an optimal policy is more taxing for the producers than the conservative or radical alternative. This is because, once the initial hike is made, agreeing on price reduction (to generate the Pindyck optimal price path) requires willingness to incur revenue losses after it has been demonstrated that unilateral price hikes can be made. Attaining agreement on a reduction in prices (or production) is undoubtedly more difficult than coordinating policies for implementing price increases.

The radical policy places the least strains on production. This policy has the greatest resource-saving effect on the exporting countries' deposits and recoverable oil-in-place relative to other tax policies. Recoverable oil-in-place is considerably higher than the reference—218.44 billion barrels by 2000 compared to 168.30 billion barrels. Of course the discovery rate is much lower. By 2000 the figures are 1,869 million barrels for the radical case, and 7,036.7 million for the reference case. By 2000 actual production is about half that of the reference case—6.1 billion barrels in comparison with 11.6 billion barrels. The differences in (radical) production between 1970 and 2000 is only about 2 billion barrels. The decline rate is undeniably lower than in the reference case. In 1977 the decline rate is identical in both cases; ten years later the radical case is at 1.4 percent in comparison with the 2.4 percent for the reference case. On all counts the oil production variables register less production, depletion, and decline—greater conservation of oil resources. This means also that there will be considerable excess capacity.

All tax policies examined create a notable decline in recoverable oil-in-place over time. The differences lie in magnitudes: how much is produced under different circumstances and what will be the relative strains on reserves and on production capacity. Production is closely tied—with appropriate adjustments lags—to the responses of the consumer and the investments of the oil companies.

Since the consumer countries have the option of increasing their own production of oil, subject to the constraints of cost and the depletion of domestic reserves, different tax policies provide different motivations for the consumer countries to draw on domestic sources of oil. But there are fundamental reserve constraints that impede the expansion of domestic production. The time paths of consumer oil production generated by the six different tax rates differ in the respective magnitudes of production mainly during the years 1980 to 1985. These years reveal the limited maneuverability of the consumer countries. A no-OPEC situation provides little incentive for expanding domestic production, since oil is cheap and there is relatively unlimited access to foreign supplies. The reference and the Pindyck optimal cases provide more incentive. The critical fact, however, is that the range between the highest amount of oil produced domestically and the lowest amount is only 2.6 billion barrels in 1987. This means that the consumers cannot readily use their own resources as a reliable political leverage vis-à-vis the producers. The optimal tax generates by necessity a higher price per barrel during these years and therefore a greater consumer response to expand domestic production. This policy entails critical political implications for the consumer countries. Adjusting to import prices by increasing domestic production touches the nerve of consumer country vulnerability. In the OECD countries where oil is available, such as in the United States, domestic oil policy can no longer be made—symbolically at least—independent of policies toward other energy sources. Deregulating prices on domestic oil is one thing, developing incentives for domestic exploitation is another, and always the constraints on known reserves lurk in the background of public debates.

The lower the tax rate is, the lower the price and the higher will be the quantity of oil demanded. The most dramatic effect is in a no-OPEC situation where the growth of demand far exceeds the reference case with attendant growth in the expected demand for Gulf oil. Total demand increases sharply during the 1980s and 1990s, and by the year 2000 it is about 13 billion barrels more than in the reference case, or a total in excess of 34 billion barrels. The breakup case has an expected demand for oil that will grow at a rate faster than the reference case toward the end of the period; almost 5 billion barrels more than the reference case are required.

With a higher oil price demand adjustments will occur. Even with the conservative policy by the end of the 1980s market adjustments result in a notable drop in demand. This adjustment will, however, be contravened by the consumers' inability to sustain permanently that decline, and a marginal recovery in demand ensues. The drop in demand is due to the price elasticity of demand and the effects of substitutes. Higher tax rates push toward alternative sources of energy. The growth in demand for all but the conservative and radical tax policies reflect the strength of domestic constraints on oil production. When the price becomes sufficiently high, demand will indeed drop sharply—at least relative to the other tax scenarios. The radical policy generates a decline in demand in the mid-1990s, following a slight increase. This shift represents the forced adjustments by the consumer countries. Demand in 1990 is close to the demand of 1970. By 2000 there is only 1 billion barrels more demanded by the consumers. This stability of demand is induced—indeed forced—by the annual increase in price stipulated in the radical tax policy.

The model-generated demand projections are lower than Pindyck's own estimates by 2.2 billion barrels for 1975 and by less than 1 billion barrels in 1980 and 1985, and from 1990 on the optimal policy estimates are about 0.5 billion barrels higher. Pindyck's own model demand for oil from the cartel is, of course, higher—by an average of 4 billion barrels throughout the period—than the model-generated demand for Gulf oil. The model produces 4.2 billion barrels in 1970 and 10.5 billion in 2000, which is almost identical to 4.2 billion barrels in 1970 and 11.6 billion in 2000 for the reference case. Figure 12.2 presents trends in consumer oil demand for the six tax policies.

Constraining—or reducing—demand must be an important aspect of any consumer energy policy. But because its effectiveness is conditioned by limitations in domestic resources and of course by institutional and political factors, it cannot be the only pillar for an overall policy. The importance of these results is in their indication of the range of demand adjustments. If prices were to increase further, consumers would of course adjust. But the oil industry will not diminish greatly in size by 2000.

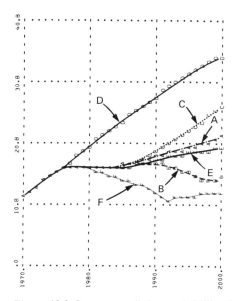

Figure 12.2 Consumer oil demand (billion barrels). A = reference case, B = conservative policy, C = breakup 1985, D = no-OPEC, E = optimal pricing, F = radical policy

12.3 Future Price of Oil

The future paths for the reference case and the five policy alternatives, in figure 12.3, represent final prices for crude oil. If the tax rate were the only determinant of price, then there would be no change in the price per barrel over time after 1970, providing the tax rate is held constant. Even in the no-OPEC situation prices increase, however marginally. This is due to increases in the cost of production and markup.

Since the consumers' response to a no-OPEC rate is to increase imports, production, the decline rate, and the rate of oil discoveries will go up. The increase in oil discoveries depletes the underlying deposits enough to drive up discovery costs to the point where they comprise 27 percent of the price increase by the year 2000. The tax rate reflecting a breakup of OPEC in 1985 generates an increased demand and higher imports and eventually will affect price.

The conservative case yields a sharp rise in 1985 and then only a marginal increase in price to 2000. These increases are due to greater production costs and the supply and demand adjustments to the change in the tax rate.

The Pindyck optimal tax case also generates a price that is for some years lower than current prices but remains higher than the reference case

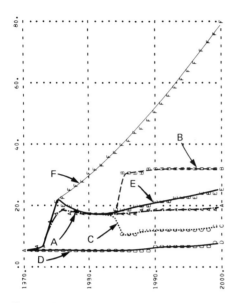

Figure 12.3 Alternative price paths (1979 dollars). A = reference case, B = conservative policy, C = breakup 1985, D = no-OPEC, E = optimal pricing, F = radical policy

throughout the 1990s. Later in this chapter we shall argue that the Pindyck price, detrimental to the consumer and oil companies, is not necessarily optimal for the producer countries' overall concerns.

The radical policy generates a price structure that exceeds the others by far—both in level and rate of increase. Most of this growth is due to the annual growth rate in the tax rate. However, there is some marginal increase in costs, but it is very small relative to the tax rate. The increase is no more than $75 in real (1979) terms from 1970 to 2000. Corporate markup increases by only a few cents in real terms since the volume of trade is constrained, thereby avoiding the necessity of expanding production and coming up against the decline rate.

The importance of markup varies from case to case. Since the markup specification in the model is an analytical device to represent the role of the companies and therefore varies under different market conditions, it would be wrong to assume that the companies' markup is always marginal. In the reference case, for instance, it is 42 percent of price in 1970, when the tax rate is $.85. By 1974 markup drops to 18 percent of price since the tax increase resulted in a substantial push in price. By 2000 markup is thirteen percent of final price. The Pindyck optimal generates a markup similar to the reference case, though lower by a few cents from 1980 on. In the no-OPEC case markup can be as high as 53 percent of price for 1975, dropping to 36 percent by 2000. The tax rate is so small that the influence of market factors is a dominant component of price. For the conservative tax rate, markup drops from 42 percent of price in 1975 to 8 percent by 2000. This contrasts with the radical policy where markup declines to 3 percent by 2000. These differences in the importance of markup are critical since they reveal the relative influence of the companies in each case. See table 12.3.

Changes in prices involve adjustments in power relations. Higher prices mean greater producer impacts on international economic transactions and, as has been demonstrated since 1973, a greater role in international politics. Claims on the resources of, and investments in, the consumer countries by necessity represent a form of political influence. Except for the embargo in 1973 to 1974 that influence has not been forcefully exerted, yet the potential remains. As the November 1979 events in Iran have shown, producer assets in consumer economies can also be used for political pressure by the West. The tax policies that have the greatest impact on the consumer countries also produce the greatest vulnerability for the producers due to the actions of the consumers.

12.4 Consumer Countries

The gains and losses to the consumer countries associated with the different tax policies are summarized in table 12.4.

Oil Demand and Oil Imports

The consumer demand for oil imports is highly responsive to the tax rate. Lower tax rates result in lower prices which increase the demand for oil and for imports, since imports are cheaper, and reduce the incentives for production from domestic sources. The three lowest tax rates exhibit the highest levels and growth in demands for imports. Since there are relatively fewer price-related incentives for expanding domestic sources of production, domestic reserves are depleted less rapidly, and their level remains higher than in the reference case. But, because the size of these reserves is not extensive to begin with, the long-term differences are not very large. Contrast these trends with the conservative and the radical tax policies. In both cases demand adjusts downward, and domestic production is increased.

The conservative policy will result in considerably less oil being imported by the consumer countries, but at a higher price. In 2000, 7.1 billion barrels will be imported from the Gulf. In real terms slightly more is spent than in the reference case. By 1985 the difference is $75 billion in

Table 12.3
Scenarios of markup as a percentage of price

Time	Reference case	Conservative	Breakup	No-OPEC	Optimal (Pindyck)	Radical
1970	42.2	42.2	42.2	42.2	42.2	42.2
1971	43.9	43.9	43.9	45.9	43.9	43.9
1972	38.6	38.6	38.6	48.3	38.6	38.6
1973	40.8	40.8	40.8	50.4	40.8	40.8
1974	17.7	17.7	17.7	52.6	17.7	17.7
1975	14.8	14.8	14.8	53.0	11.8	11.8
1976	13.5	13.5	13.5	51.9	12.1	10.7
1977	13.3	13.3	13.3	52.0	12.2	9.5
1978	13.4	13.4	13.4	52.4	12.2	8.4
1979	12.9	12.9	12.9	51.4	12.2	7.4
1980	12.5	12.5	12.5	50.4	12.4	6.7
1981	12.4	12.4	12.4	49.6	12.2	6.1
1982	12.5	12.5	12.5	48.7	12.1	5.7
1983	12.6	12.6	12.6	47.7	12.2	5.3
1984	12.8	12.8	12.8	46.8	12.3	5.0
1985	13.0	7.4	20.9	45.8	12.3	4.7
1986	13.2	7.5	21.1	45.0	12.3	4.5
1987	13.3	7.6	21.3	44.1	12.3	4.3
1988	13.6	7.6	21.8	43.3	12.4	4.5
1989	13.9	7.7	22.4	42.4	12.5	4.6
1990	14.0	7.6	22.3	41.6	12.3	4.5
1991	13.9	7.6	22.2	41.1	12.1	4.1
1992	13.8	7.5	22.0	40.6	11.8	4.0
1993	13.7	7.4	21.8	40.0	11.5	4.3
1994	13.6	7.3	21.6	39.3	11.3	4.2
1995	13.5	7.3	21.4	38.6	11.0	4.0
1996	13.4	7.2	21.2	38.0	10.8	3.8
1997	13.3	7.0	21.0	37.4	10.5	3.6
1998	13.2	7.0	20.8	36.9	10.3	3.4
1999	13.1	7.3	20.6	36.5	10.1	3.2
2000	13.0	7.5	20.4	36.0	9.9	3.0

real terms. But imports must increase. By 2000 the difference is only about $16 billion. The effects are more pronounced during the years 1985 to 1990, when higher tax rates cost the consumers an average of $75 billion in import payments relative to the reference case.

The radical policy illustrates best the limits on the consumers' ability to cut imports. Between 1975 and 1980 there is a clear 1 billion barrel cut; by 1990 another 3 billion barrels are taken off the import schedule. But the downward trend cannot be sustained. There is a gradual increase in import demand, reaching 9.5 billion barrels in 2000—which is roughly the level of imports for 1979. Thus the radical tax policy would affect the imports of the consumer countries significantly, especially when compared to the reference case which registers 18.7 billion barrels of imported oil in 2000.

Consumer oil import demand for the Pindyck optimal case is similar to the reference case. In 1970, 6.9 billion barrels of imports were demanded, and by 1990 the figure was 11.7 billion barrels for the Pindyck optimal tax case and 12.1 billion in the reference case. By the end of the period the reference case generates 2 billion barrels more oil imported than in the optimal case—but the price is different.

These two types of responses generate the eventual leveling off of demand observed in figure 12.2 during the decade following the tax increase. Because of the constraints on consumer domestic reserves and the depletion of these reserves, domestic production cannot be sustained at high levels, and the result is a marginal recovery or relative increase in demand for imports by the last years of the period.

Managing domestic demand for oil is a major policy issue in the consumer countries. Lower rates do not eliminate internal political problems for the importing countries but skew them away from the idiom of cost to that of security. Higher demand, met by imports, is increasing the fears of dependence in both the public and policy makers. Dependence is the corollary of weakness. Shortage has become a deep concern, but security supersedes all these issues as oil prices change. The worldwide debate pertaining to the nature of demand and its management is increasingly political. There is no policy position on this issue that will not automatically trigger formidable opposition.

Balance of Payments

Consumer oil import expenditures indicate the financial stress on the consumer countries generated by their imports. The net effects of different tax policies are shaped both by the price per barrel and the amount imported. Figure 12.4 presents the oil bill generated by the different tax

Table 12.4
Consumer countries' forecasts for the year 2000 with the effects of alternative policies

Policy variable	Initial value (1970)	Reference case
	(billion barrels)	
Consumer oil reserves	200.00	47.64
Consumer oil demand	11.14	21.03
Consumer oil production	4.2	2.31
Consumer oil import demand	6.94	18.71
	(billion 1979 dollars)	
Consumer oil import expenditures	20.91	216.97
Discounted consumer oil import expenditures[a]	20.29	50.31
Discounted consumer oil import expenditures cumulated	0	1,608.7
Consumer basic balance	−6.92	−127.93
Consumer basic balance cumulated	0	−1,737.9
Discounted consumer basic balance	−6.92	−29.83
Discounted consumer basic balance cumulated	0	−719.98
Consumer balance of trade	−15.47	−181.32
Discounted consumer balance of trade	−15.47	−42.04
Discounted consumer balance of trade cumulated	0	−1,379.2
Consumer current account	−6.92	−163.74
Discounted consumer current account	−6.92	−37.97
Discounted consumer current account cumulated	0	−1,169.5

Conservative policy	Breakup 1985	No-OPEC	Optimal (Pindyck)	Radical policy
44.66	49.74	63.37	46.67	38.73
14.48	25.75	34.26	19.19	12.13
2.53	2.14	2.72	2.37	2.62
11.96	23.61	31.54	16.82	9.51
232.55	181.68	144.40	248.69	453.87
53.92	42.13	33.48	57.67	105.24
1,902.8	1,400.3	865.8	1,720.2	2,391.8
− 118.33	− 129.71	− 126.38	− 127.42	− 137.70
− 1,755.5	− 1,602.2	− 1,159.9	− 1,749.8	− 1,818.7
− 27.58	− 30.26	− 29.61	− 29.69	− 32.14
− 729.30	− 665.37	− 430.66	− 724.09	− 756.92
− 191.74	− 150.65	− 122.71	− 210.37	− 400.65
− 44.46	− 34.93	− 28.45	− 48.78	− 92.90
− 1,653.5	− 1,183.3	− 705.87	− 1,483.5	− 2,108.1
− 179.33	− 130.50	− 112.32	− 194.22	− 390.34
− 41.58	− 30.26	− 26.04	− 45.04	− 90.51
− 1,459.1	− 965.70	− 426.74	− 1,279.4	− 1,945.7

Table 12.4 (continued)

Policy variable	Initial value (1970)	Reference case
Consumer capital account	0	35.11
Discounted consumer capital account	0	8.14
Discounted consumer capital account cumulated	0	450.52
Values of remaining reserves[b]	1,003.6	923.0
Discounted value of remaining reserves	1,003.6	214.0
Discounted value of remaining reserves cumulative	0	30,906.0

[a] The present discounted value is $PDV + Y_t/(1 + r)^t$, where PDV is the value in year zero to yield Y in year t, assuming an interest rated r.
[b] Computed as follows:
Consumer value of remaining reserves = Consumer remaining reserves · Price.

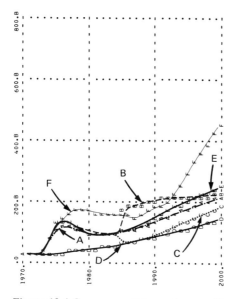

Figure 12.4 Consumer oil import expenditures (billion 1979 dollars). $A =$ reference case, $B =$ conservative policy, $C =$ breakup 1985, $D =$ no-OPEC, $E =$ optimal pricing, $F =$ radical policy

Conservative policy	Breakup 1985	No-OPEC	Optimal (Pindyck)	Radical policy
60.37	0	− 15.39	66.17	251.72
14.00	0	− 3.57	15.34	58.37
729.81	300.32	− 3.92	555.25	1,188.8
1,453.5	640.1	484.8	1,151.9	3,092.1
337.0	148.4	112.4	267.1	717.0
36,158.0	28,234.0	12,350.0	32,945.0	53,751.0

rates. In general all tax policies result in higher oil payments by the consumer countries, but for different reasons. The conservative policy of doubling tax rates in 1985 results predictably in a higher oil import bill computed over the entire period. The jump in 1985 is therefore characteristic of the consumers' inability to adjust rapidly to higher prices. The slight decline in real terms during the mid-1990s is due to reduction in the volume of consumer imports, and the subsequent rise is because of their inability to sustain the cuts. These trends reflect high prices more than the volume of imports. The reference case, which presumes a continuation of the 1978 tax rate to 2000, also yields a steady growth in oil payments due, after 1978, largely to the growth in imports.

The Pindyck optimal price generates an import bill even greater than the conservative OPEC policy by the end of the period. In 1970 the consumers spent $20.9 billion in constant dollars, and in 1975 their import bill jumps to $132 billion, while their imports from the Gulf increase from 4.2 billion barrels to 6.2 billion barrels. With a subsequent drop in price, and then a gradual recovery, the same pattern appears in the oil import payments of the consumer countries. By 2000 the expenditures reach the $248 billion mark because the price under a Pindyck optimal tax policy is higher than in the reference case.

The radical policy also creates a very high bill, by far the highest of the alternative policy runs. The gains to the consumers can be gleaned from their capital account. The initial surge in producer foreign investments, which appear as credits in the capital account of the consumers, is after the oil price increase of 1973. The next ten years reflect dramatic growth in the capital account. The subsequent leveling off in the early mid-1980s is due to the adjustments of the consumers to higher oil prices and the producers' own responses to these adjustments. Thus by the late 1980s there is decline in the consumers' capital account. But since oil revenues grow and are directed into foreign investments by the producer countries, there is a phenomenal increase in the last decade of the period. By 2000 the capital account registers $252 billion in 1979 dollars, in comparison with $35 billion for the reference case and $66 billion in the Pindyck optimal.

Of course one must argue that these gains are offset to a large extent by the trends in the balance of trade and in the current account. The basic balance registers strong negative trends throughout the whole period; however, the differences are not dramatic when compared to the reference case which holds the tax rate constant from 1978. The differences are imperceptible until the final year of the period. By 2000 the radical case registers only $10 billion deficit larger than the reference. Despite a sharp downward trend in demand the import bill is so high due to price per barrel that demand cannot adjust downward sufficiently to affect imports substantially enough that the import bill will not skyrocket.

Paradoxically the no-OPEC tax policy also creates a substantial oil bill, but largely because of the amount imported. With lower prices demand rises, imports grow, and the oil bill responds accordingly. In real terms the consumers spent about $20.9 billion in 1970 on their oil imports from the Gulf, and this increased by $16 billion by 1978. By the end of the period import payments for that year are over $144 billion in real terms in comparison with $217 billion in the reference case.

Clearly lower prices are more advantageous for the consumer countries if gains are calculated solely in terms of oil payments and their implications for the balance of trade. (See figure 12.5 for trends in the consumer current account.) But lower taxes influence directly the foreign investments of the producer countries and are reflected in the capital account of the consumers. There are long-term trade-offs and some real costs. In the no-OPEC case the capital account registers a zero balance for most of the period, and is negative in the closing years. Thus the capital account offsets none of the deficits engendered by the current account, unlike the reference case. No-OPEC would generate a marginally better current account, with a $51 billion difference in real terms for the terminal year. The basic balance is improved over the reference case in the no-OPEC case by only $1 billion for 2000. The reason is that although expenditures on oil create a slightly less unfavorable situation by comparison with the

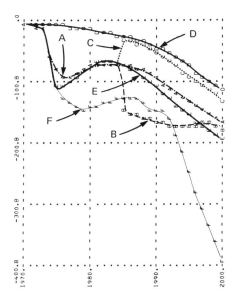

Figure 12.5 Consumer current account (billion 1979 dollars). A = reference case. B = conservative policy, C = breakup 1985, D = no-OPEC, E = optimal pricing, F = radical policy

reference case, the inability of the producers to invest abroad represents a major loss for the consumer countries. (See figure 12.6 for the capital account.)

For the cumulative balance of payments—a special creature of this model—it appears that by the end of the period the deficit of the consumer countries will exceed the $1,000 billion mark in 1979 dollars for every tax policy. In more immediate terms, however, the deficit, as registered by the basic balance, is least unfavorable for the consumer countries under no-OPEC tax conditions. During the 1970s, the deficit is at about $8 billion annually, but it subsequently deteriorates rapidly. For ten years or so, however, the consumers' basic balance remains less than $20 billion in constant prices. Still in the longer run the cumulative balance of payments position in the no-OPEC case is much more favorable than in the reference case due to low prices and low imports in the early years of the period. In the absence of producer foreign investments, a lower import bill for the consumer countries mitigates the balance of payments predicament. Even so these trends differ from those in the other scenarios only in the absolute levels involved.

Perhaps ironically, by the end of the period a conservative tax policy generates only $88.2 billion more in terms of consumer oil payments than the no-OPEC case. This is not a small amount, but it is far smaller than

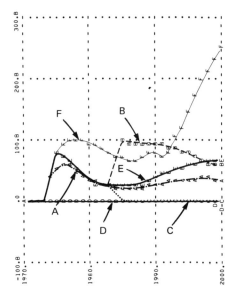

Figure 12.6 Consumer capital account (billion 1979 dollars). A = reference case, B = conservative policy, C = breakup 1985, D = no-OPEC, E = optimal pricing, F = radical policy

the relative differences in price. The radical policy of course costs the consumers $154.3 billion in 1985, $183.4 billion by 1990, and $454 billion in 2000 (all in 1979 dollars).

Every tax policy from the most benign to the optimal from the producers' perspective generates a large, growing oil bill for the consumer countries. In the case of the low tax rate policy this growth is mainly due to the volume imported; in the case of higher taxes per barrel it is due to the price of oil. In both types of situation the result is an expanding oil bill for the consumers, but the largest variations are in the amounts of capital inflows associated with the investments of the producer countries. Cutting oil taxes is not a solution to the consumer countries' payments predicament.

A forceful stance against the producer countries would be politically popular in the West, particularly in the United States, but its effects are more symbolic than practical. The political problem is as follows: with or without OPEC the consumers will pay more for oil—in one case because of higher prices, in the other because of more imports. Attempting to convince the public of this basic fact amounts to stepping in a quagmire. In the absence of an energy policy or a strategy for meeting imports, rationing supplies, or exploiting domestic resources, it is politically expedient to place blame on the producers. The real political problem lies at home:

finding ways of meeting demand without incurring additional costs or without running into a variety of constraints—many resulting from previous policies, or lack of policy.

Strategic Vulnerability

Paradoxically the consumers' strategic vulnerability tied to the volume of oil imported will be greatest in a no-OPEC situation. It will also be high in the event of an OPEC breakup. This is because there are relatively fewer constraints on demand in comparison with the other cases, imports from the Gulf soar, and there are no incentives for investments in energy alternatives since the price of oil is so low. The position of least vulnerability in the long run is that generated by a radical policy. Vulnerability declines as demand is lowered, especially for Gulf imports, and substitute energy resources are utilized. The comparative trends are in figure 12.7.

While on economic grounds the no-OPEC case may appear to be the most favorable, the political implications are extensive. The underlying vulnerability makes of the consumer a potential political hostage. Reliance upon foreign petroleum is a possible source of weakness. The absence of OPEC assumes the political weakness of the producers internationally. It was only a matter of time before collusion among the pro-

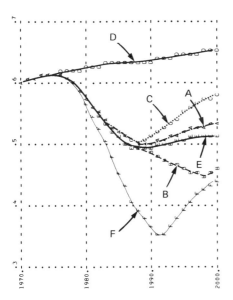

Figure 12.7 Strategic vulnerability. A = reference case, B = conservative policy, C = breakup 1985, D = no-OPEC, E = optimal pricing, F = radical policy

ducers occurred. Low taxes create illusionary security for the consumers—one that is sound economically but fraught with pitfalls politically. History has vindicated these observations. Collusion among producers evolved in large part because of awareness of the economic and political inequities implied by low taxes.

From the consumer countries' viewpoint therefore an increase in the tax rate per barrel will result in the decline of overall long-term vulnerability. The different tax policies will begin demonstrating their effects by the early 1980s, with a substantial divergence in the vulnerability. A lower tax rate will invariably accentuate and reinforce the consumers' vulnerability. An increase will improve this position. The conservative case will contribute to the improvement of the importing countries' strategic position. The difference is only marginal when compared with the reference case, but, given the sharp increase in taxes in 1985, this result seems paradoxical because the consumers are prompted to cut their imports.

An improvement in the consumers' strategic vulnerability will entail political costs domestically but political benefits internationally. Adjusting to severe external constraints on energy supplies will require retrenchment of demand and policies designed to regulate and enforce that retrenchment. It will also involve the exploitation of domestic reserves, at least over a certain number of years and therefore a clear, not incremental, change in domestic oil policies. In both cases the government's stance will inevitably mobilize public opposition.

A decline in the reliance on foreign sources involves a decline in susceptibility to political leverage from these quarters. There are some political costs to the consumer countries, however, since a strong action such as a radical tax policy would reaffirm their inability to influence dramatically the world market or the producers' policies.

A radical tax policy will result in a strong increase in substitute energies. With increased prices alternative sources of energy will become competitive economically. The Pindyck optimal and the reference case reveal a similar, though weaker, response by the end of the period. The dominant influences on substitute availability of course are the price of oil and the falling costs of substitutes. This is why, even in the no-OPEC case, there is a notable, but gradual, increase in alternative energies. (See figure 12.8.)

As substitutes become economically viable, the political benefits for consumer countries will be considerable. Since they continue to control sophisticated energy technologies, the buyers in the oil market will be sellers in alternative energy markets. The oil producers of today are

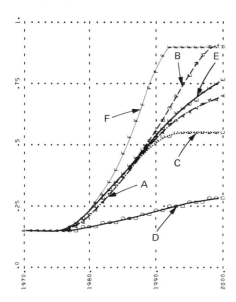

Figure 12.8 Substitute availability. A = reference case, B = conservative policy. C = breakup 1985, D = no-OPEC, E = optimal pricing, F = radical policy

cognizant of this eventuality, and the First Arab Energy Conference, March 1979, reaffirmed the importance of access to technology.

The emerging rhetoric among oil producers and others is to talk of an exchange context, where raw material will be made available at just prices if technological advances in the energy and other fields were made available to them. The nature of the exchange remains vague. The interests are clear, but the terms of the exchange are not specified.

The index of insecurity of supplies, part of the strategic vulnerability of the consumers, reflects the importance of Gulf imports for meeting demand. With lower oil prices there will be higher demand and more oil imports. Thus the no-OPEC tax policy represents these dual effects in a sharp increase in the insecurity variable. The lower the price of oil, the greater will be the consumers' draw upon insecure sources. The decline in insecurity evidenced for the Pindyck optimal, the conservative case, the radical policy, and the reference case is due to the drop in demand as the initial response to the price hikes. Since the consumers cannot sustain this cut, as demand rises, they have no option but to meet this rise with Gulf imports. The uniform increase in supply insecurity exhibited by the six tax policies—at different levels and with varying rates—is due to this inability to constrain demand or meet it from other than Gulf sources. (See figure 12.9.)

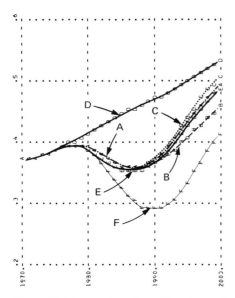

Figure 12.9 Insecurity of supplies. A = reference case, B = conservative policy, C = breakup 1985, D = no-OPEC, E = optimal pricing, F = radical policy

From a political perspective the security of Gulf sources is becoming increasingly critical. Cuts in production, shifts in trade patterns, and intervention in regulated market processes cannot be ruled out. The revolution in Iran has made regional implications more serious than ever before. We do not argue that Saudi Arabia or Kuwait will also be subjected to similar revolutionary pressures, but that they are more likely to use their production schedules as a policy instrument for political purposes as well, rather than as a means of sustaining prevailing prices. The insecurity of supplies index is only a rough indicator and cannot address itself to these contingencies; however, it does reveal the context within which these issues may become more or less salient on economic grounds.

The same processes, and attendant responses, are found in a third component of vulnerability, namely, imports relative to consumption. In this case, however, the consumers' inability to increase domestic production sufficiently to meet their import requirements contributes to such a dramatic growth of this indicator over time. The declines for the reference case, the Pindyck optimal, and the two other higher tax rate policies during the decade of the 1980s are due to the initial consumer response of cutting import demand. Since such cuts cannot be sustained, because of both economic growth and limitations in production from domestic sources, the import/consumption ratio shoots up in all six cases.

In sum the no-OPEC case creates the greatest costs of strategic vulnerability when all its components are taken into account. Lower prices will mean greater imports and therefore a higher extent of vulnerability. The availability of substitutes, most evident in the conservative, the radical, and the Pindyck optimal policies, is the major offsetting factor. The resulting flexibility will create a new international political environment with a structure and allocation of power very different from the case where the consumer nations do not have the oil substitutes.

Domestic Oil Reserves
Higher domestic production will seriously strain the volume of domestic reserves remaining in 2000. The political issues are whose revenues will be depleted first and faster and what could be done to protect the value of remaining reserves. Perceptions of these issues will undoubtedly color policies. The no-OPEC scenario registers the greatest benefits for the consumer countries since, as they import most of their requirements, the decline in reserves is less pronounced than in the other tax scenarios. However, then the value of the remaining reserves will be lowest due to low, prevailing world prices. Assessing the existing reserves in prevailing prices at every point in time, and discounting to the present, yields a lower valuation of reserves in the no-OPEC case than in any of the other scenarios.

Domestic oil policy is already an important political issue for all consumer countries. For those that have sizable reserves the issue will be posed in terms of exploitation rate and governmental interventions in the process. For those that do not, the issues will be posed in terms of various sharing schemes and access to the production of other consumers. Invariably there will be greater interdependence in the political debates of the consumer countries themselves. Developing a set of internally consistent domestic policies remains high in the political agenda of consumer countries.

12.5 Producer Countries

The different tax rate policies generate fundamentally different income streams for the producer countries. The highest revenues over time are those associated with both a conservative and a more radical policy. Revenues shoot up dramatically and, though reductions in imports attempted by the consumer countries result in a temporary fall in revenues, they remain at a higher level than in any other situations; the recovery during the last decades of the period results in additional growth in the income

Table 12.5
Producer countries' forecasts for the year 2000 with effects of alternative policies

Policy variable	Initial value (1970)	Reference case
	(billion barrels)	
Producer undiscovered oil	500.00	427.59
Producer recoverable oil-in-place	350.00	168.3
Producer oil discovery rate	1.05	7.04
Producer oil production	4.28	11.63
	(billion 1979 dollars)	
Producer industrial capital	134.6	898.52
Producer investments	0	1,972.3
Producer imports	5.44	35.65
Producer oil income	9.55	147.50
Discounted producer oil income[a]	9.55	34.20
Discounted producer oil income cumulated	0	1,170.9
Producer foreign investments	0	35.11
Discounted producer foreign investments	0	8.14
Discounted producer foreign investments cumulative	0	450.52
Producer investment income	0	98.61
Discounted producer investment income	0	22.87
Discounted producer investment income cumulative	0	388.08
Value of remaining reserves[b]	3,011.0	7,404.0
Discounted value of remaining reserves	3,011.0	1,716.7

Conservative policy	Breakup 1985	No-OPEC	Optimal (Pindyck)	Radical policy
439.33	418.72	298.66	431.92	460.70
189.11	155.46	107.66	174.26	218.44
3.70	9.84	28.33	6.17	1.87
7.59	14.55	19.27	10.50	6.15
1.007.0	824.62	617.06	934.60	1,152.7
3.184.4	1,322.0	0	2,426.2	5,184.4
40.81	31.03	21.70	38.32	53.23
188.57	92.49	16.93	186.94	419.00
43.72	21.45	3.93	43.35	97.16
1.512.4	932.03	161.16	1,298.0	2,079.2
60.37	0	0	66.17	251.72
14.00	0	0	15.34	58.37
729.81	300.32	0	555.25	1,188.8
159.22	66.10	0	121.36	259.2
36.92	15.33	0	28.14	60.11
523.92	331.83	0	443.01	826.8
13.304.0	4,695.0	1,961.2	9,632.0	35,829.0
3.084.9	1,088.6	454.8	2,233.0	8,308.0

Table 12.5 (continued)

Policy variable	Initial value (1970)	Reference case
Discounted value of remaining reserves cumulated	0	124,260.0

[a] The present discounted value is $PDV = Y_t/(1 + r)^t$, where PDV is the value in year zero to yield Y in year t, assuming an interest rated r.
[b] Producer values of remaining reserves = [(Producer undiscovered oil
                    · Oil recovery fraction)
                    + Producer recoverable oil-in-place] · Price.

stream. Table 12.5 presents the summary results from the perspective of the producer countries.

Oil Revenue

The radical tax policy yields extensive and unparalleled oil revenues. The growth takes place after the mid-1970s. The jump in 1975 is twofold over the previous year's revenue. Because of the consumers' inability to curtail demand over time, or to expand production, and because both the price per barrel and the number of barrels imported grow sharply during the last decade of the period, the producers are able to obtain oil revenues over three times greater than the reference case, and over twice the Pindyck optimal case. The present discounted value of future income streams is $2.08 trillion in the radical case, in comparison with $1.17 trillion for the reference and $1.30 trillion for the optimal case. Even the conservative policy yields higher income streams—$1.51 trillion.

Although the Pindyck optimal policy generates marked increases in revenue over the early decades, this policy is clearly dominated by the radical-OPEC and in some years by the reference case as well. The OPEC breakup yields income streams substantially lower than the three higher tax rate policies. The no-OPEC situation, with a real decline in the tax rate over time, creates the lowest revenues for the producer countries. Figure 12.10 presents the annual levels of producer oil revenue associated with the different tax policies. From the discounted value of future revenue it is clear that the conservative and radical policies still dominate all the other alternatives in the longer run. The Pindyck optimal revenue stream is not as great as in the radical case.[3]

In terms of the present discounted value of the cumulated income stream for the oil-exporting countries under different tax policies, note that both the radical and the conservative tax policies dominate the Pindyck optimal price policy by nearly $781 and $214 billion, respectively, in

Conservative policy	Breakup 1985	No-OPEC	Optimal (Pindyck)	Radical policy
156,450.0	108,370.0	43,342.0	135,010.0	259,130.0

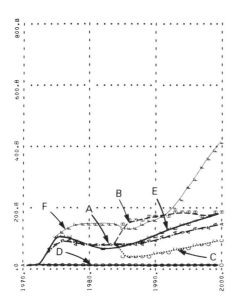

Figure 12.10 Producer oil income (billion 1979 dollars). A = reference case, B = conservative policy, C = breakup 1985, D = no-OPEC, E = optimal pricing, F = radical policy

1979 dollars. However, the Pindyck optimal generates $127 billion more in revenues for the producer countries than in the reference case where the 1978 tax rates remain constant to the end of the period.

These results demonstrate that the Pindyck optimal policy does not achieve its goals for the producer countries in this model. Rather in terms of domestic capital investments the radical policy dominates. On the other hand, when present discounted values of future investment streams are taken into account, the Pindyck optimal dominates the reference case for most of the period. The same can be said with regard to imports of goods and services. Note the differences in implications for the producers' internal investments and their imports of goods and services. Parenthetically it should be recalled that the producer oil consumption patterns associated with different tax policies are very similar to each other.

Producer Investments

The total investments of the producer countries are derived from additions to investment that include foreign investment plus earnings returns from their previous investments in the economies of the consumer countries. As stipulated in the model, producer investments are initialized at zero in 1970, and the subsequent growth is generated entirely endogenously. Figure 12.11 presents the trends in producer investments. A radical policy will yield the most extensive total investments. In the revenue streams growth is the most dramatic during the final decade of the period. The net effect, however, is for a dramatic increase over time. By 2000 the producer investments in the radical policy case surpass the $5,180 billion mark. The Pindyck optimal is considerably less favorable.

The foreign investments of the producer countries are affected fundamentally by the differences in tax policies and associated revenues. Recall that their foreign investments derive from the difference between oil revenues and the amounts spent on domestic investments and imports of goods and services. In those terms the foreign investment stream is a residual and will reflect both revenue and the other development allocations. The drastic reduction in income in the no-OPEC case has inevitable consequences for development variables. Import demand is less than two-thirds in value in comparison with the reference case, and demand

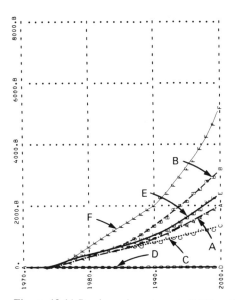

Figure 12.11 Producer investments (billion 1979 dollars). A = reference case, B = conservative policy, C = breakup 1985, D = no-OPEC, E = optimal pricing. F = radical policy

for domestic investment is a little more than one-half the value in the reference case. There is some growth, since there is a growth in income and in the other explanatory variables (population and capital stock), but the levels are very low. By the end of the period import demand in real terms would be at the 1978 level of the reference case, and investment demand would be at about the 1980 level.

In the OPEC breakup situation there are also strong adverse effects. In 1985 foreign investments drop from $23 billion in the previous year to zero—remaining at zero for the rest of the period. Contrast this with the reference case where foreign investments grow to $35 billion by 2000.

In the radical case from the 1980s onward the level of foreign investments of the producer countries is dramatic, well over twice the amounts in the reference case for the 1980s and over three times for the 1990s. During the last decade of the period allocation to foreign investments shoot up totally unrestrained. This explosiveness reflects the growth in oil income and the constraints in the other areas for absorbing revenue.

Producer Oil Reserves
The conservative and radical policies create greater volume and value of remaining resources for the producer countries and dominate every other policy option. They entail a lower rate of oil production, a lower discovery rate, higher remaining reserves in 2000, and greater underlying deposits. Likewise the present discounted value of remaining reserves is greatest; they dominate the Pindyck optimal by $852 and $6,075 billion, respectively.

The no-OPEC case would obviously generate the least benefits for the producer countries. Oil production is higher, the discovery rate is the steepest, and recoverable oil-in-place the lowest by 2000. The value of the remaining reserves will be about one-sixth of the value in the radical case and one-fourth of the value for the Pindyck optimal. With a no-OPEC revenue stream there simply are no surplus revenues for external investment. The producers' total investments—foreign investment plus investment additions due to income earned on earlier investments—remain at zero throughout the period. (This result has direct implications for the long-term capital inflows in the balance of payments of the consumer countries.)

12.6 Corporate Profits and Oil Investments

Over the long term the oil companies would increase their profits by almost sixfold in a no-OPEC situation. This is a case in which oil demand

Table 12.6
Corporate forecasts for the year 2000 with effects of alternative policies (billion 1979 dollars)

Policy variable	Initial value (1970)	Reference case
Oil capital	2.69	32.58
Profits	8.82	28.17
Discounted profits	8.82	6.53
Discounted profits cumulative	0	244.52
Repatriated profits	8.54	17.58
Discounted repatriated profits	8.54	4.08
Discounted repatriated profits cumulative	0	209.65
Total investments	.27	10.60
Costs	2.47	39.88
Discounted costs	2.47	9.25
Discounted costs cumulative	0	187.12

and oil trade grow dramatically over time. Profits would increase in real terms from $8.8 billion in 1970 to over $51 billion by the end of the period. Growth of oil demand induces an adjustment in supply and motivates the oil companies to expand their capability for production through investments in exploration and development. Higher demand invariably gives the companies a greater role in the management of the industry. With higher demand will come greater investment in exploration and development which will directly affect oil capital and production capacity. Over time exploration investments exceed development investments by about $40 billion (in real terms for 2000) in contrast to an initial condition in 1970 that generated much higher development than exploration investments. This reflects the increasing strain on recoverable oil-in-place and ultimately on the underlying deposits.

Any increase in the tax rate per barrel results in a decline in corporate profits relative to the no-OPEC case. Since such profits are derived from markup, they depend on the volume of trade; the increase in the tax per barrel cuts into the corporations' profits if the consumers' response to higher oil prices is to reduce their imports. Markup itself responds to market conditions affected by the price of oil. These influences are indirect,

Conservative policy	Breakup 1985	No-OPEC	Optimal (Pindyck)	Radical policy
19.50	39.56	52.58	28.83	16.11
17.39	37.05	51.93	24.60	13.83
4.03	8.59	12.04	5.71	3.21
220.34	259.84	383.99	236.25	180.29
12.41	20.15	10.39	16.15	10.31
2.88	4.67	2.41	3.74	2.39
194.40	217.63	279.12	204.11	162.31
4.98	16.90	56.93	8.45	3.52
25.12	50.76	74.17	35.71	19.52
5.82	11.77	17.20	8.28	4.53
163.72	202.34	314.68	179.65	125.82

but they eventually bear upon corporate profits. Table 12.6 summarizes the effects of different tax policies for the corporations.

The conservative situation generates the lowest corporate profits compared with the other tax rate policies (except the radical case). On balance the companies will have less profits beyond 1985 than in the reference case. The differences are about $11 billion in real terms for 2000, which represents a drop of 39 percent. Again this is due to the fact that a higher tax rate means greater consumer expenditures, resulting in attempts to cut imports. The Pindyck optimal price generates corporate profits very similar to those of the reference case. Profits do not exhibit strong fluctuation. The trend is steady and increases from 1985 to 2000 by $12 billion in real terms. This tax policy is of course not optimal from the companies' perspective—nor was it intended to be—since any tax policies that stipulate a higher tax rate will generate less oil trade and therefore smaller profits for the companies.

Corporate profits sharply decline in the decade of the 1980s due to higher costs. By 2000 the radical tax policy enables the companies to reap only a little less than half of the profits of the reference case—$14 billion in comparison to $28 billion in 1979 dollars. The investment allocations are substantially higher for development than for exploration, both proportionally and relative to the reference case. These trends are due to adjustments in oil sales resulting from consumer efforts to cut demand.

Figure 12.12 presents the comparative trend in corporate profits generated by the different tax policies. Note the behavior of profits under the conservative and radical situations. The declines in the mid-1990s are due to consumer cuts in imports that cannot be sustained and the subsequent rise in profits to the upswing in imports. The political role of the companies internationally will be tied to their economic strength, and profit is a good, though imperfect, indicator of this strength.

The corporations' repatriated profits, which consist of profits less investments, generally follow a trend similar to overall profits. By the last decade of the period, however, the corporations greatly increase exploration investments in the no-OPEC case, which generates a sharp drop in repatriated profits to the year 2000.

Different tax rate policies will also affect corporate costs of production in approximately the same way. At this writing average costs are a relatively small fraction of price, and only over the long run, as production must be expanded, oil discovery costs increase notably. Production costs, however, increase by a factor of six throughout the period. In sum, al-

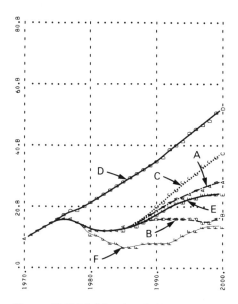

Figure 12.12 Multinational oil profits (billion 1979 dollars). A = reference case, B = conservative policy, C = breakup 1985, D = no-OPEC, E = optimal pricing, F = radical policy

though the effects of alternative tax policies on costs are not dramatic, costs do in fact grow over time. Corporate production costs are financed from oil profits. Costs go up because they are a function of exports times basic costs factors (the discovery cost and the cost per unit capacity). When more is produced and imported, both types of costs go up.

Desired oil production capacity, which is a function of expected demand and desired capacity utilization, reveals the expected production requirements generated by each tax policy. Predictably the lowest tax rate contributes to the highest desired production capacity and by extension actual capacity. The no-OPEC scenario is most favorable for the companies, since it generates greater oil sales. In this limited sense the interests of the companies diverge strongly from both those of the consumer countries and the producer countries. The need for more production capacity will enhance the companies' role and possibly influence in the oil market.

The investments of the corporations in development and exploration are designed to contribute to allocation and planning in the industry. It is only in the cases of extreme pressures on reserves that the exploration investments increase dramatically due to the need to draw increasingly upon deposits. For most of the tax rate situations examined, investments in development respond most immediately to changes in the conditions of the industry, since these changes are manifested directly in demand which generates changes in production requirements. The tax rate policy that entails the least constraints on demand, hence on production, calls for the greatest investment in the oil industry.

12.7 Evaluating Prices

What do we expect for the future?

Different assumptions and modes of analysis generate different price paths. However, the comparisons in figure 12.13 are instructive. The no-OPEC price path is roughly similar, and slightly higher, than the price projected under competitive conditions by Nordhaus (1973). The conservative policy creates a price higher than any other path projected to 2000, with the exception of the radical policy. The two Ben-Shahar (1976) projections, a one-shot increase in 1982 and an initial increase plus and 8 percent per year growth rate, generate paths lower than the radical. Pindyck's own path of course is over the long run roughly similar to the reference case.

At this writing even the conservative price path appears outmoded. Yet it is higher than alternative projections by other analysts: the future holds a path between the radical and the conservative tax policies. Given both

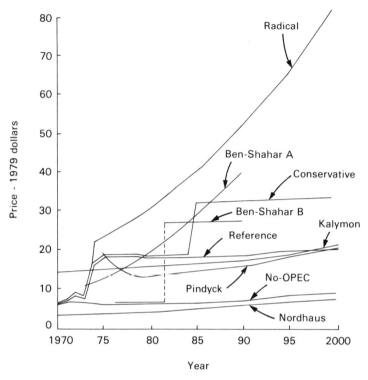

Figure 12.13 Price forecasts. Notes: Ben-Shahar A = initial price of $7 in 1974 dollars, 8 percent annual increase. Ben-Shahar B = one-shot increase from $4 to $17.65 in 1974 dollars. Pindyck price based on OPEC as a monopoly and a 5 percent discount rate. Nordhaus assumes a competitive market.

political and economic conditions, lower price paths are simply unrealistic. With the exception of the radical case, the price policies examined in this book are well within the range of other projections. At this writing the future is one of higher prices. The radical policy—while dramatic from today's perspective—may even be conservative in years to come. The cohesion of OPEC is no longer the single most critical determinant of prices. Indeed the record of the spot market for 1979 reflected progressively higher prices, although in 1980 they began to decline. Whether the spot market is an indication (a barometer) of things to come or whether it is a reflection of the extreme fluctuations in the market (an aberration) remains of considerable debate.

13
Conclusion

This book is not an argument for higher prices but an acknowledgment of the worldwide ramifications of oil prices—whatever those prices might be. The argument is that globally the energy issue is clearly an economic issue. But every economic problem has critical political and social implications. This economic issue, which is perceived differently by different parties to energy exchanges, must be placed in a broader political context. Economic criteria for evaluating the effects of alternative tax rates or alternative market conditions are inherently political. Consumers will be better off with lower prices now under certain conditions and on strictly economic grounds, but there are political reasons that lower prices might not be in their best interests.

This book is not a defense of OPEC. On economic grounds alone of course low prices are not bad for consumers. But can we argue that money now is better than money later? The low-price position can be supported only if we can assume perfect financial markets or real investments. Consumers will benefit from lower prices now only if they can convert their present benefits into future gains which will at least offset future losses.

Low prices will, however, inevitably have attendant consequences that are bad for consumers, given prevailing political values in the West. They will involve lower constraints on the growth of energy demand, greater imports from outside, potentially insecure, sources and reduced incentives for the development of alternative energy sources. Low prices will also generate a deterioration in the balance of payments, due to the volume of imports. That this deterioration will be less severe than the deficits with higher prices in no way invalidates the basic point: greater oil imports—at whatever prices—will involve greater net outflow of real financial resources, unless these outflows are offset by exports of goods and services and by capital inflows.

Low prices are of course not desirable for oil-exporting countries. Aside from the obvious costs associated with lower tax rates—accelerated production, greater demand for exports, and reduced oil revenues—there are secondary costs. These involve limitations on resources for domestic investment and, more fundamentally from the viewpoint of the consumer countries, limitations on the financial resources for investments in the economies of the oil-consuming countries. Low prices erode one important basis for offsetting the balance of payments deficits of the consumer countries incurred under a scenario of lower oil prices.

Higher oil prices therefore are obviously in the best interest of the oil-exporting countries. Revenues are higher, investments are greater, production is lower, and remaining reserves are greatest. Higher prices will contribute to the conservation of reserves. The value of remaining oil in the ground for the producer countries will be greater than with lower prices. Higher prices will also accentuate the economic interdependence among producer and consumer countries.

The profits of the oil companies accrued from oil sales are assured. With low or high prices the volume of oil trade generates profits for the companies. Profits are higher with low prices, since the volume of imports is greater. Higher prices result in reduced demand, thereby constraining corporate profits. Higher oil prices also reduce the companies' role in evaluating, even managing, the worldwide industry's conditions. The markup device of our analysis is peculiar from an economist's point of view. Yet it represents a key adjustment process in the world oil system. With low prices markup can be as high as 50 percent of price. As the price rises, the proportion of markup to final price drops sharply.

We have specified the indicators used to make judgments regarding costs and benefits of different tax rates. For instance, lower prices, which are better on stricly economic grounds, will contribute to stronger expressions of vulnerability and dependence on the part of consumer countries. These political considerations are extremely important in determining what the consumer governments will do in response to oil prices. The strategic vulnerability indicators reveal the implications of alternative tax rates for the consumer countries and confirm that the importers are less vulnerable with higher tax rates.

The skeptical reader might argue that none of this makes much sense. The fact is that much of the political debate revolving around energy issues entails fear of vulnerability. Project Independence, promoted in the United States in 1974, was predicated on that sentiment. Even its label was a testimonial to this fear. The fact that it never amounted to much economically does not negate the underlying sentiment of fear of dependence that any political analyst, journalist, or layman would recognize.

We do not argue that vulnerability is inevitable or that there are no conceivable policies to counter dependence on external sources of energy. We do argue that with lower prices there is greater vulnerability and thus more dependence. At this writing President Carter's energy posture retains little economic content, considerable political salience, and a large dose of emotional frustration and dissatisfaction. These latter considerations are peripheral to economic analysis.

This book is framed in the idiom of economic analysis. The IPE model takes the producers' key policy—the tax rate—as an input and traces its ramifications. The emphasis is on the oil production processes, market conditions, and petroleum trade. But every economic indicator entails fundamental political implications. The inquiry into the implications of alternative oil prices is a political analysis. There are numerous, possibly peripheral, issues in this book that point to the interdependence of politics and economics. For example, higher tax rates generate higher revenues for the oil-producing countries, greater surplus income, and greater investments in the economics of the consumer countries. Investments from OPEC are clearly not the only, or best, way of buttressing the consumers' balance of payments. They may well entail important political costs and accentuate fear of dependence on the oil-producing countries. But they are a part of the economic reality of higher oil prices, and they must be acknowledged accordingly. This book does not argue that in the absence of collusion among the oil-producing countries prices will be too low to clear the market—hence a bad situation economically. It does argue that low prices do have certain important, negative consequences.

Clearly there is nothing intrinsically good or bad about high or low demand for oil. There are only perceptions, costs, and implications economically and politically. From an analytical perspective alternative levels of oil demand entail political implications, if this demand is met by imports perceived as insecure. This is not our own perception but one commonly expressed in the public arena in the West, and therefore it cannot be ignored.

So too one may or may not be alarmed at the prospects of oil running out. Different tax rates will induce different rates of production, depletion, and investments in expanding reserves. In the long run petroleum is a depletable resource. However, this is not the issue. The economics of depletable resources are well elucidated. The timing associated with alternative production rates is at issue, as is the relationship to different tax rates. It does make a difference for the producers—and possibly the consumers—when oil will run out and whose oil will run out.

Alarm is not the intent of this book. We do not lament nor belabor the exhaustion of a depletable resource—others have done that more poignantly and thoroughly. This book shows the interconnections among the actions and reactions of the different agents in the world oil market and the effects of the market itself on the broader set of relations among nations.

What can be done? How strong are these results? What are the effects of different demand- and supply-related policies?

For the consumers to place less reliance on external sources at least three conditions must hold: (1) expansion of domestic reserves and production, (2) increased responsiveness of oil demand to price, and (3) expansion of non-Gulf sources of imports.

This book assumes consumer reserves to be about 200 billion barrels (in 1970).[1] An alternative estimate of ultimate discovery reserves of 250 billion barrels, including the U.S. and North Sea reserves is also reasonable.[2] There is considerable uncertainty about the demand elasticity. Some long-run estimates are as high as 0.9 for $12 per barrel. This book assumed 0.3. Consumer domestic production is estimated on the basis of initial OECD estimates. By raising the consumer initial reserves to 250 billion, increasing the elasticity of demand to 0.5, and freeing domestic production from the constraints of the OECD assumptions, the first two conditions are met.

The third condition—expanding non-Gulf imports—can be met by allowing non-Gulf production to respond to price and depletion effects, subject of course to the constraints of production possibilities. By setting non-Gulf reserves to 600 billion barrels, with a price elasticity of 0.1 (for comparison purposes of 0.3), the potential aggressiveness of non-Gulf suppliers can be roughly gauged.[3] Given the dominant role of government policy in developing countries, the actual elasticity is extremely difficult to determine.

For a no-OPEC situation, none of these interventions will have any effects. A world with no OPEC would be essentially as we have described it in chapter 12. Of course the incentives for consumer policies are negligible. Imports expand, with non-Gulf supply being substantially lower so that the position of strategic vulnerability actually worsens (by 10 percent relative to the basic no-OPEC scenario). This is due to vastly greater reliance on insecure Gulf supplies, bringing a situation of exhaustion in that area by the end of the period. The projected figure of 75 million bbl/day for the year 2000 is so unrealistic by today's standard that the upward pressures on price seem greater than previously demonstrated. Of course noneconomic reasons for such a tax rate, effective consumer pressures and so on, could have created such a situation.

By contrast strong consumer measures in the radical case would have dramatic effects. If the reserves of the consumer countries were 25 percent greater than stipulated, domestic production will expand, generating a drop in the Gulf production to below 3 billion barrels. This means that in some years substantial production cutbacks must occur and some agreements must be reached among the producers to determine how this will be done and who will bear the burden.

A higher demand elasticity (0.5 as opposed to 0.3) will have sharp effects on oil demand. By 2000 demand can be cut by one-third, relative to the base radical scenario. The present discounted value of the producers' oil revenue would be cut by 33 percent. At one point oil production in the Gulf could hit a dramatic low of 2 billion barrels.

If the consumer countries expanded their reserves and increased their adjustments to oil prices, their Gulf production could drop below 1 billion barrels for as long as a three-year stretch. Oil revenues would be driven down virtually to the reference case.

Non-Gulf sources of supply play a critical role. If these were allowed to be entirely price responsive, their impact in the radical scenario is dramatic. Non-Gulf resources expand and exacerbate the effects of consumer oil demand reduction on Gulf production. This combined policy—conditions (1) and (2)—generates a strong leverage on the producers.[4] This could even impose a sustained cut in production.

But what would happen if the Gulf producers preempted this response and imposed a shutdown?

Testing this situation counterfactually—by simulating the Iran cutbacks in 1978 and 1979—entails the loss of 1 billion barrels of production capacity in the last quarter of 1978, 1.5 billion barrels in the first quarter of 1979, 1 billion barrels in the second and third quarters of 1979, and 0.5 billion barrels in the fourth quarter. In a vein of optimism we stipulate a return of all production capacity by the first quarter of 1980.

In a no-OPEC world, where the market is tighter than in other scenarios, the effects of this shutdown would be severe. The difference between the consumers' oil demand and the amount of oil imported reaches a high of 1.6 billion barrels. This means that this amount desired from the Gulf is simply not met by Gulf sources. Corporate markup under these conditions increases by 20 percent, to 72 percent of final price. The price of course will be pushed up largely by the companies' markup. This reflects the tightness of the market. Essentially the oil companies increase their investments and productions from existing production capacity, inducing a fairly rapid adjustment in the market.

In the radical policy case the effects of production shutdown are marginal because excess capacity exists. The market is relatively loose, and reactions are milder and of shorter duration. The point is that a loss of production capacity in a competitive market would have a substantial—negative—impact. In a world with a radical OPEC the necessary dramatic adjustments would have already occurred so that added loss of capacity would not have notable effects. The role of the companies is critical in both situations because they engineer the market adjustments.

An essential factor is excess capacity. Imports can be met from accelerated production of existing sources and the eventual expansion of productive capacity from these sources. In cases where there is little excess capacity, as in the no-OPEC situation, this latitude does not exist, and the loss of capacity exerts a severe strain on the system internationally. Ironically it is in the radical case that the effects of production cutbacks will be most readily adjusted to by the market as a whole.

Again this is not to advocate shutdown policies but to stress their potential effects. In the long run the international system cannot be expected to be infinitely resilient. Everyone must adjust and bear the burdens of straining an already stressed system.

Postscript

Sharp oil price increases of December 1979 and the adjustments of 1980 raise yet another issue: OPEC prices remain below the prices quoted at the Rotterdam market. The spot market represents a higher price structure than has been imposed by OPEC. Figure 13.1 presents trends in both OPEC and spot transactions. Spot prices react more rapidly to market conditions, whereas OPEC pricing reduces short-term fluctuations.[5]

In 1973 OPEC was regarded as a major disrupter of the world oil market. This view persisted for the next five years. Yet during this time a change in the producer countries' role in the market was taking place. This change remained unnoticed until the explosion of oil prices at the Rotterdam exchange throughout 1979. The oil producers' official prices remained far below those quoted on spot exchanges. Today, far from being a disrupter of the market, the oil-exporting countries appear to be regulators. Their price preferences are conservative: while participating to some extent in spot transactions, OPEC's formal position is to oppose the action of the Rotterdam exchange in setting prices and prevent spot transactions from dominating the oil market.

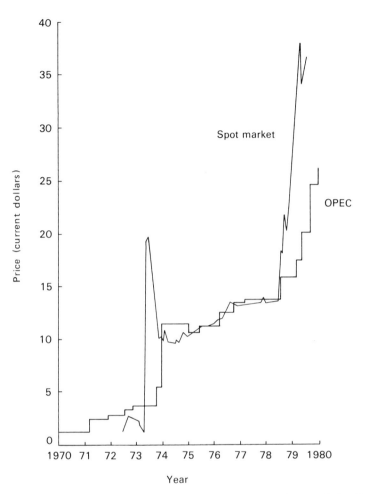

Figure 13.1 Comparison of OPEC and spot market price trends for Saudi light. Historically OPEC has used Saudi light for its market crude, although at present a substantial portion of OPEC crude is not tied to the Saudi light.

The final irony of course is that OPEC, at this writing a conservative regulator of the international fuel exchanges, regards the spot market fever as an aberration that undermines its own strength in oil trade. The more appropriate, current role of OPEC is as a stabilizer. Whether OPEC can now effectively set prices appears in doubt. Rather the strength of the Rotterdam market and the persistence of the upward trend is sustained by the producers' production policies and periodic announcements of cutbacks.

Further whether the spot market can continue to set prices in the long run remains also in doubt. It might instead turn out to be a barometer of the prevailing oil market or of trends to come. In either case spot transactions signal the evolution of the market. New strains are being felt. New adjustments are required, and new policy responses must be adopted by all agents in the oil market—the producer countries, consumer countries, and international oil companies.

The producer countries' expansion into downstream operations adds another strong element to the market. It generates new sources of potential stability and integration of producer investments in the oil industry worldwide. This integration will further strengthen interdependence in the world oil market and the exchanges generated by prevailing conditions at any point in time.

Appendix A
List of Variables and Units

Table A.1
Units of measurement of the supply sector

Variable	Variable name	Unit of measurement
PUO	Undiscovered oil	Barrels
PODR	Discovery rate	Barrels/year
EOEI	Effective oil exploration investment	$/year
PDT1	Production delay time constant 1	Years[a]
PODC	Oil discovery cost	$/barrel
PUOF	Undiscovered oil fraction	Percentage
PUOIP	Unrecoverable oil-in-place	Barrels
PROIP	Recoverable oil-in-place	Barrels
ORF1	Recovery fraction	Percentage
POP	Oil production	Barrels/year
POCU	Capacity utilization	Percentage
POPC	Production capacity	Barrels/year
POC	Oil production capital	$
EODI	Effective oil development investment	$/year
PDT2	Producer delay time constant 2	Years[b]
ODIR	Oil development investment rate	$/year
POCD	Oil capital depreciation	$/year
PALOC1	Average lifetime of capital	Years
POPDR	Production decline rate	Percentage
POPCUC	Production cost per unit of capacity	$/barrel
TGOD	Total Gulf oil demand	Barrels/year
PC	Producer states' consumption	Barrels/year
CGOM	Consumer gulf oil imports	Barrels/year
NPOPC	New production capacity	Barrels/year
POPCDA	Production capacity loss to depreciation	Barrels/year

[a] Producer delay time constants are labeled 1 and 2 to enable a differentiation between them.
[b] PDT2 is the second producer delay time constant in this sector. This labeling procedure is common practice in system dynamics models. See Pugh (1976).

Table A.2
Units of measurement of the finance sector

Variable	Variable name	Unit of measurement
COME	Consumer oil import expenditures	$/year
POT	Producer oil taxes and royalties	$/year
POY	Producer oil income	$/year
MOC	MNC oil production costs[a]	$/year
MOP	MNC oil profits	$/year
MPARI	MNC profits available for reinvestment	$/year
PRIF1	Profit reinvestment maximum	Percentage
MRIOP	MNC reinvested oil profits	$/year
MRPOP	MNC repatriated oil profits	$/year
TOI	Total oil investment	$/year
ODI	Oil development investment	$/year
OEI	Oil exploration investment	$/year

[a] MNC refers to international oil companies.

Table A.3
Units of measurement of the management sector

Variable	Variable name	Unit of measurement
DPOCU	Desired capacity utilization	Percentage
ED	Expected demand	Barrels/year
MT1	Management time constant 1	Years
DPOPC	Desired production capacity	Barrels/year
DPOPCA	Desired additional production capacity	Barrels/year
POPCD	Production capacity loss	Barrels/year
DDID	Desired development investment discrepancy	$/year
DDI	Desired development investment	$/year
MAF2	Management adjustment fraction 2	Percentage
DPOPDR	Desired production decline rate	Percentage
DDIS	Desired discoveries	Barrels/year
DEI	Desired exploration investment	$/year
DI	Desired investment	$/year
MDEF	MNC development/exploration fraction	Percentage
EPOI	Expected producer oil investment	$/year
DMI	Desired MNC investment	$/year
MDOI	MNC direct investments	$/year

Table A.4
Units of measurement of the price sector

Variable	Variable name	Unit of measurement
PMOC	Cost of production	$/barrel
P	Price	$/barrel
MNCM	MNC markup on price	$/barrel
POPDRM	Production decline rate multiplier	1.0 + percentage
COMGM	Consumer oil import gap multiplier	1.0 + percentage
CM1	1970 oil markup	$
POCUD	Capacity utilization discrepancy	Percentage

Table A.5
Units of measurement of the producer sector

Variable	Variable name	Unit of measurement
PPOP	Producer population	Individuals
PPOPG1	Population growth rate	Percentage
PK	Producer states' industrial capital	$
PKD	Producer industrial capital depreciation	$/year
PKDR1	Average lifetime of industrial capital	Years
PKID	Capital investment demand	$/year
PIIR1	Producer constant impact on investment demand	Dimensionless (regression estimate)
PKIR1	Producer industrial capital impact on investment demand	Dimensionless (regression estimate)
PYIR1	Producer oil income impact on investment demand	Dimensionless (regression estimate)
PPIR1	Producer population impact on investment demand	Dimensionless (regression estimate)
POYL	Producer oil income (lagged value)	$/year
PAT1	Producer adjustment time 1	Years
PCD	Producer oil demand	Barrels/year
PICR1	Producer constant impact on resource demand	Dimensionless (regression estimate)
PPCR1	Producer population impact on resource demand	Dimensionless (regression estimate)
PKCR1	Producer industrial capital impact on resource demand	Dimensionless (regression estimate)
PTMD	Producer total import demand	$/year
PIMR1	Producer constant impact on import demand	Dimensionless (regression estimate)
PCMR1	Producer resource consumption impact on consumer demand	Dimensionless (regression estimate)

Table A.5 (continued)

Variable	Variable name	Unit of measurement
PKMR1	Producer industrial capital impact on import demand	Dimensionless (regression estimate)
PYMR1	Producer oil income impact on import demand	Dimensionless (regression estimate)
POTR	Producer oil tax rate	$/barrel
POI	Producer oil investment	$/year
PKI	Producer industrial capital investment	$/year
PTM	Producer total imports	$/year
PRIY	Producer repatriated foreign investment income	$/year
PFI	Producer foreign investments	$/year
PIR	Producer investments remitted	$/year

Table A.6
Units of measurement of the consumer sector

Variable	Variable name	Unit of measurement
COSP	Substitute price (all substitutes converted to oil equivalents)	$/barrel
COSA	Substitute availability	Dimensionless
CPESA1	Price elasticity on substitute availability	Dimensionless
CAT1	Consumer adjustment time 1	Years
COP	Consumer oil production	Barrels/year
CDPL	Depletion effect on production	Percentage
COR	Consumer oil reserves	Barrels
COPR	Consumer oil production rate	Barrels/year
CORRF	Consumer oil reserve fraction remaining	Percentage
COD	Consumer oil demand	Barrels/year
COSAM	Substitute availability multiplier on demand	Percentage
COSAM1	Elasticity on substitute availability	Dimensionless
CPMOP	Price multiplier on production	Dimensionless
CP1	Base price for comparison	$/barrel
CPEOP1	Price elasticity on production	Dimensionless
CAT2	Consumer adjustment time 2	Years
CPMOD	Price multiplier on demand	Dimensionless
CPEOD1	Price elasticity on demand	Dimensionless
CAT3	Consumer adjustment time 3	Years
COMD	Consumer oil import demand	Barrels/year
CGOMD	Consumer Gulf oil import demand	Barrels/year

Table A.6 (continued)

Variable	Variable name	Unit of measurement
PGEF	Persian Gulf export fraction	Percentage
COMCR	Consumer import consumption ratio	Percentage
COSD	Consumer supply/demand ratio	Percentage
COMG	Consumer import shortfall	Barrels/year
EP	Expected price	$/barrel
CAT4	Consumer adjustment time 4	Years
COE	Consumer oil essentiality	Percentage
COIS	Consumer oil supply insecurity	Percentage
CT1	Time constant 1	Years
COSV	Consumer strategic vulnerability	Percentage
CR	Interest rate	Percentage
COPT	Consumer oil production table	Barrels/year
CODS	Consumer oil demand series	Barrels/year

Table A.7
Units of measurement of the international economic sector

Variable	Variable name	Unit of measurement
CBT	Consumer balance of trade	$/year
CBCA	Consumer balance on current account	$/year
CBKA	Consumer balance on capital account	$/year
CBB	Consumer basic balance	$/year
CBBC	Consumer cumulative basic balance	$
PI	Producer foreign investments	$
PIA	Producer additional foreign investments	$/year
PIRR	Producer foreign investment remittance rate	Percentage
PRY	Producer reinvested income from foreign investment	$/year
PIY	Producer foreign investment income	$/year

Appendix B
Documentation

Table B.1
Supply sector

Variable type	Variable label	Definition
Initial value	PUO	Undiscovered oil (1970)
Initial value	PUOIP	Unrecoverable oil-in-place (1970)
Initial value	PROIP	Recoverable oil-in-place (1970)
Initial value[a]	POC	Oil capital stock (1970)
Time delay	PDT1	Delay 1
Time delay	PDT2	Delay 2
Constant	ORF1	Oil recovery fraction
Table function	POPCUCT	Oil production cost per unit of capacity
Constant	PALOC1	Average lifetime of capital
Table function	PODC	Oil discovery cost

Note: C = complete documentation; P = partial documentation. The finance sector does not require documentation since all variables are endogenous.
[a] In 1970 dollars.

Comment/description	Value	Documentation	Source
Estimates on total oil-in-place (recoverable and nonrecoverable) not yet discovered for Gulf states	500 billion barrels	C	(12)
Gulf state oil-in-place not recoverable with 1970 technology and price	450 billion barrels	C	(3) p. 71
Gulf states recoverable oil-in-place with 1970 technology and price	350 billion barrels	C	(13)
1970 U.S. dollar value of oil capital equipment in Gulf states	$1,000 billion	P	(16) p. 12
Period of delay between the moment an investment is made in exploration and the moment that investment affects the oil discovery rate	5 years	C	(1)
Delay between the moment an investment is made in field development and the moment in which the new capital is operational in the field. Eight months is normal.	$2\frac{1}{2}$ months	C	(1)
Percent of newly discovered oil recoverable with 1970 technology	50 percent	C	(3) pp. 71, 32
Capitalized cost of production	See table B.7	C	(2)
Conservative estimate, based on the rate of depreciation for oil capital equipment	10 years	C	(1)
Per barrel cost of discovering a barrel of oil. Table is constructed as a function of the percentage of undiscovered oil still undiscovered and assumes that, the more the Gulf's ultimate reserves are located, the higher will be the cost of additional discoveries.	See table B.7	C	(3) pp. 7, 74, 76

Table B.2
Management sector

Variable type	Variable label	Definition
Constant	DPOCU1	Desired capacity utilization
Time delay	MT1	MNC demand expectation horizon
Coefficient	MAF2	MNC adjustment fraction 2; replacing the value of oil capital lost through depreciation in last period
Constant	DPOPDR1	Desired production decline rate

Table B.3
Price sector

Variable type	Variable label	Definition
Initial value	CM1	1970 Oil markup

Comment/description	Value	Documentation	Source
Represents the percentage of full capacity that the producers would like to be producing at a particular moment	0.90	C	(20) p. 26
Related to the smooth on expected demand	1 year		
If the amount of capital depreciation on the last period is greater than the desired development investment discrepancy, then the desired development investment will be (MAF2) percent of the amount of capital when depreciated; otherwise DDI will equal DDID.	0.80	P	
Estimates of decline rate in the United States range between 8 and 14 percent. In the Gulf actual decline rates of 1.3 percent are observed. Desired decline rate is usually 6.7 percent; 10 percent indicates conservative exploration policy.	0.10	C	(1)

Comment/description	Value	Documentation	Source
Historically determined per barrel base markup for Saudi light crude; equals 1970 price minus costs and tax rate	0.82	C	(3) (7) p. 15 (9)

Table B.4
Consumer sector

Variable type	Variable label	Definition
Exponent	CPESA1	Elasticity on substitute availability
Time delay	CAT1	Consumer adjustment time on substitute availability
Constant[a]	CP1	Base price for comparison
Exponent	CPEOP1	Consumer supply elasticity
Table function	COET	Consumer essentiality of oil
Time delay	CT1	Supply security time constant
Constant	CR1	Interest rate

Comment/description	Value	Documentation	Source
The exponent (the elasticity) on the substitute availability equation where $$\text{Substitute availability} = \left(\frac{\text{Expected oil price}}{\text{Price of substitute energy}}\right)^{\text{CPESA1}}.$$ (See comments on COSAM1.)	0.85	P	
The adjustment time associated with availability of substitute: it would take about 10 years to bring a new energy on line.	10 years	P	(18) p. 8 (19) pp. xvi, xvii
A base price against which the current expected price is compared in determining the price multiplier on production for the consumer states, namely, how world price will affect the decision on oil production in the consuming state: $$\text{Price multiplier on production} = \text{Delay}_{3}\left\{\left(\frac{\text{Expected price}}{\text{CPI}}\right)^{\text{CPEOP1}}, \text{CAT2}\right\}.$$	$1.80	C	(7) p. 15
Elasticity (exponent) on the expected price/base price function which determines the short-term response of domestic oil supply to change in Gulf oil price (see equation above for functional form), assumed that consumer oil production is relatively inelastic in the short run	0.3	C	(6) pp. 366–367
Soft, intuitive measure of the importance of petroleum to the consumer economy. This variable is used in the strategic vulnerability equation. See chapter 8.	See table B.7		(4), (15)
Period of time in a smooth function over which the exponential average of supply security is taken: $$\text{Supply security} = \text{Smooth}\left(\frac{\text{Imports}}{\text{Import demand}}, \text{CT1}\right).$$	4	P	
Given the time horizon of the model, and the fact that it is specified to generate output in constant 1970 dollars, a long-term interest rate of 5 percent is used.	0.05	P	

Table B.4 (continued)

Variable type	Variable label	Definition
Initial value	CORO	Consumer oil reserves
Table function	CDPLT	Depletion effect
Time delay	CAT2	Consumer adjustment time on domestic production
Exponent	CPEOD1	Consumer demand elasticity for oil
Time delay	CAT3	Consumer demand adjustment time
Time delay	CAT4	Consumer adjustment time for expected price
Initial value [a]	EP1	Initial expected price (1970)
Table function	COSPT	Substitute price (as a function of time)

Comment/description	Value	Documentation	Source
Rough estimate of ultimate recoverable reserves	200 billion barrels	C	(12)
Represents the effects of oil depletion upon domestic production table. Values are dependent upon the fraction of remaining reserves; as reserves are depleted, production is inhibited. This is a variable in the consumer oil production equation.	See table B.7		
As part of a third-order information delay this constant represents an average response time for domestic production to adjust to world price (more specifically to Gulf oil price)	5 years	C	(22) p. 79
This is part of the equation measuring the response of consumer nation demand to the world Gulf oil price:	0.3	C	(6) pp. 366–367
Price multiplier on demand $= \mathrm{Delay}_3 \left\{ \left(\dfrac{\mathrm{CP1}}{\mathrm{Expected\ price}} \right)^{\mathrm{CPEOD1}} , \ \mathrm{CAT3} \right\} .$			
As part of a delay the value represents (roughly) an average time for consumer states to adjust their level of demand	5 years	C	(14) vol. 2, p. 11
This time constant has a slightly different interpretation than the adjustment times associated with delays. The expected price function is an exponential average of the observed price for the past X years. Here $X = 3$. Expected price = smooth (price, CAT4).	3 years	P	
The world (Gulf) oil price in 1970, $1.68 to $2.55	$1.80	C	(7), p. 15
This is the price per barrel of energy substitutes at ten-year intervals, 1970 to 2000.	See table B.7	P	(17) p. 19

Table B.4 (continued)

Variable type	Variable label	Definition
Table function	COPT	Consumer oil production (as a function of time)
Table function	CODT	Consumer oil demand (as a function of time)
Coefficient	PGEF	Persian Gulf share of world petroleum exports
Exponent	COSAM1	Elasticity for substitute availability multiplier

a In 1970 dollars.

Table B.5
Producer sector

Variable type	Variable label	Definition
Initial value	PPOP1	Producer population in 1970
Exponent	PPOPG1	Producer population growth rate
Constant	PKDR1	Producer average lifetime of capital

Comment/description	Value	Documentation	Source
Number of barrels of oil produced in the relevant consumer states 1970 to 2000, based upon OECD projections which assume 1970 price. After 1985 production is assumed to climb for five more years and then stabilize. In the functional form of the production equation these figures are then modified by the price multiplier on production.	See table B.7	C	(14) pp. 47, 50, 53
Same as COPT in form and also based upon OECD projections which assume 1970 price. These projections are then extrapolated linearly from 1985 to 2000.	See table B.7	C	(14) pp. 47, 50, 53
Annual data for world crude petroleum exports for 1970 to 1972 yields Gulf share of 52 to 58 percent of world exports; selection of 60 percent based on assumption of a persisting world share by the six Gulf states. See chapter 8.	0.6	C	(4), (15), (5)
Determines the sensitivity of consumer oil demand (COD) to the availability of petroleum substitutes. In test runs an elasticity of 0.2 resulted in a moderate impact upon oil demand by substitute availability, comparable to other studies.	0.2	P	(10)

Comment/description	Value	Documentation	Source
This coefficient is an initialization value, but in the functional form of the producer population variable it winds up as a constant. The 47 million figure represents the population for Gulf states in 1970.	47 million	C	(21)
Documentation for Gulf states revealed a range of growth rates between 1.8 and 5.8 percent; 3.1 percent used as a weighted mean	0.031	C	(21)
Form: $PKD.KL = \dfrac{PK.K}{PKDR1}$.	30 years	P	(8)

Table B.5 (continued)

Variable type	Variable label	Definition
Constant	PIIR1	Producer capital investment demand constant
Coefficient	PKIR1	Producer capital coefficient on investment demand
Coefficient	PYIR1	Producer income coefficient on investment demand
Coefficient	PPIR1	Producer population coefficient on investment demand
Coefficient	PAT1	Producer adjustment time 1
Coefficient	PICR1	Producer oil consumption constant
Coefficient	PPCR1	Producer population coefficient on oil consumption
Coefficient	PKCR1	Producer capital coefficient on oil consumption
Coefficient	PIMR1	Producer import demand constant
Coefficient	PCMR1	Producer population coefficient on import demand
Coefficient	PKMR1	Capital coefficient on import demand
Coefficient	PYMR1	Producer income coefficient on import demand
Initialization[a]	PKO	Producer non-oil industrial capital stock in 1970
Coefficient	PRIY1	Producer repatriated investment income constant
Table function	POHTRT	Producer oil historical tax rate table
Table function	POPTRT	Producer oil policy tax rate table
Table function	POIT	Producer oil investment table

[a] In 1970 dollars.

Comment/description	Value	Documentation	Source
Regression estimate	−0.598	C	(11), (14), (21)
Regression estimate	0.058	C	(11), (14), (21)
Regression estimate	0.104	C	(11), (14), (21)
Regression estimate	32.96	C	(11), (14), (21)
This is part of a smooth equation (POYL) in the investment demand equation. POYL is the exponential average of past oil income (PAT1) specifying an averaging process that takes the past 2 years into account.	2 years		
Regression estimate	−0.35	C	(11), (14), (21)
Regression estimate	1.95	C	(11), (14), (21)
Regression estimate	0.0005	C	(11), (14), (21)
Regression estimate	1.239	C	(11), (14), (21)
Regression estimate	7.67	C	(11), (14), (21)
Regression estimate	0.03	C	(11), (14), (21)
Regression estimate	0.045	C	(11), (14), (21)
Level of capital stock (in dollars) in Gulf states in 1970 based on all non-oil industries. Estimate is derived from figures on GDP.	$50 billion	P	(11) pp. 203, 237, 319, 421
Percentage of producer investments abroad repatriated in a given period; set at zero indicating no repatriation	0	P	
Runs from January 1970 to July 1978 at six-month increments, in 1970 dollars	See table B.7	C	(9) (3) p. 208
Used to test different tax policies	See table B.7		
Oil investment by Gulf states (in dollars), 1970 to 2000	See table B.7	P	

Table B.6
International economic sector

Variable type	Variable label	Definition
Initial value	CBBC0	Consumers basic balance cumulative in 1970
Initial value	PI0	Producer investments in 1970

Comment/description	Value	Documentation	Source
Cumulative capital and current balance between consumer and Gulf states in 1970.	0	na	na
Level of dollars invested by producers in 1970. This is not a measure of the investments Gulf states made in 1970 but a measure of the outstanding investments held by the producers to which new investments will be added and remittances subtracted: it is a stock not a flow.	0	na	na

Table B.7
Table values

Time series running from 1970 to 2000, with data points at five-year intervals

POPCUCT[a] = 0.2/0.365/0.535/0.7/0.87/1.035/1.2,

COET   = 0.55/0.533/0.516/0.5/0.483/0.468/0.45,

COSPT[a]   = 20/15/12/10/10/10/10,

COPT   = 4.2e9/4.9e9/5.7e9/6.1e9/6.5e9/6.5e9/6.5e9,

CODT   = 11.5e9/16e9/20.5e9/25e9/30e9/35e9/40e9,

POPTRT   = 0/0/0/0/0/0/0,

POIT   = 0/0/0/0/0/0/0.

Time series running from 1970 to 1978.5, with data points at 0.5-year intervals

POHTRT[a] = 0.858/0.858/0.90/1.17/1.24/1.24/1.06/1.26/4.09/4.85/5.02/5.02/ 5.48/5.48/5.33/5.62/4.94/4.94.

Not time series but dependent on other variables

PODCT[a]   = 0.75/0.40/0.30/0.20/0.02,

CDPLT   = 0/0.25/0.75/0.9/1.

[a] In 1970 dollars.

References

(1) Adelman, M. A. 1979. Private correspondence of November 27, 1979.

(2) Adelman, M. A. 1976. The World Oil Cartel: Scarcity, Economics, and Politics. *Quarterly Review of Economics and Business,* 16: 7–18.

(3) Adelman, M. A. 1972. *The World Petroleum Market.* Baltimore: Resources for the Future, Johns Hopkins University Press.

(4) British Petroleum Company, Ltd. 1979. *Statistical Review of the World Oil Industry.* London.

(5) Central Intelligence Agency. 1977. *The International Energy Situation: Outlook to 1985.* Washington, 1977. D.C. (April).

(6) Choucri, N. 1979. Analytical Specifications of the World Oil Market. *Journal of Conflict Resolution,* 23: 346–372.

(7) DeGolyer & MacNaughton. 1975. *Twentieth Century Petroleum Statistics, 1975.* Dallas: DeGolyer & MacNaughton.

(8) Eckaus, Richard S., Ford International Professor of Economics, MIT, personal communications.

(9) Energy Economics Research Ltd. in cooperation with Middle East Economic Survey. 1979. *International Crude Oil and Product Prices.* Beirut: Middle East Petroleum and Economic Publications, pp. 3–4.

(10) Herman, Stewart W., James S. Cannon, with Alfred J. Malefatto. 1977. *Energy Futures: Industry and the New Technologies.* Cambridge, Mass.: Ballinger Publishing Co.

(11) International Monetary Fund. 1976. *International Financial Statistics.* Vol. 24 (May 1976).

(12) Moody, John D. and Robert E. Geiger. 1975. Petroleum Resources: How Much Oil and Where. *Technology Review* (March/April): 38–45.

(13) *Oil and Gas Journal.* December 28, 1970.

(14) Organization for Economic Co-operation and Development. 1974. *Energy Prospects to 1985: An Assessment of Long Term Energy Developments and Related Policies.* Vols. 1 and 2. Paris: Organization for Economic Co-operation and Development.

(15) Organization for Economic Co-operation and Development. 1977. *World Energy Outlook.* Paris.

(16) Organization of Petroleum Exporting Countries. 1974. *OPEC Bulletin, 1973.* Vienna: OPEC.

(17) Policy Study Group of the MIT Energy Laboratory. 1977. *Recent Proposals for Government Support for the Commercialization of Shale Oil: A Review and Analysis.* Energy laboratory report, May 20, 1977.

(18) Project Independence Blueprint, Final Task Force Report. 1974. *Potential Future Role of Oil Shale: Prospects and Constraints.* Washington, D.C.: Government Printing Office.

(19) Project Independence Blueprint, Final Task Force Report. 1974. *Synthetic Fuels from Coal.* Washington, D.C.: Government Printing Office.

(20) Stobaugh, R. and D. Yergin (eds.). 1979. *Energy Future: Report of the Harvard Business School Energy Project.* New York: Random House.

(21) United Nations, Department of Economic and Social Affairs. 1975. *Demographic Yearbook, 1973.* New York: United Nations.

(22) U.S. Congress, Senate, Committee on Interior and Insular Affairs. 1973. *Hearings, Oil and Gas Imports Issues.* 93rd Congress, 1st session.

(23) Workshop on Alternative Energy Strategies. 1977. *Energy: Global Prospects, 1985–2000.* New York: McGraw-Hill.

Bibliography

Abolfathi, F., G. Keynon, M. D. Hayes, L. A. Hazlewood, and R. Crain. 1977. *The OPEC Market to 1985.* Lexington, Mass.: Lexington Books.

Adelman, M. A. and J. L. Paddock. 1979. An Aggregate Model of Petroleum Production Capacity and Supply Forecasting. MITEL 79-005. Revised. Cambridge, Mass.: MIT Energy Laboratory.

Adelman, M. A. 1976. The World Oil Cartel: Scarcity, Economics and Politics. *Quarterly Review of Economics and Business,* 16: 7–18.

Adelman, M. A. 1972. *The World Petroleum Market.* Baltimore: Johns Hopkins University Press.

Akins, J. E. 1973. The Oil Crisis: This Time the Wolf Is Here. *Foreign Affairs.* 51: 462–490.

American Petroleum Institute. 1969. Forecast for the Seventies. *Oil and Gas Journal,* 10 (November): 159–204.

Ben-Shahar, H. 1976. *Oil Prices and Capital.* Lexington, Mass.: Lexington Books.

Bhattacharya, A. K. 1977. *The Myth of Petropower.* Lexington, Mass.: Lexington Books.

Blair, J. M. 1976. *The Control of Oil.* New York: Pantheon Books.

Blitzer, C., A. Meeraus, and A. Stoutjesdijk. 1973. A Dynamic Model of OPEC Trade and Production. *Journal of Development Economics,* 2: 319–335.

Bobrow, D. B., and R. T. Kudrle. 1976. Theory, Policy and Resource Cartels: The Case of OPEC. *Journal of Conflict Resolution,* 20: 3–56.

Bohi, D. R. and M. Russell. *U.S. Energy Policy: Resources for the Future.* Baltimore: Johns Hopkins University Press, 1975.

British Petroleum Company, Ltd. 1976. *BP Statistical Review of the World Oil Industry.* London.

Central Intelligence Agency. 1977. *The International Energy Situation: Outlook to 1985.* Washington, D.C. (April).

Choucri, N. 1979. Analytical Specifications of the World Oil Market: A Review and Comparison of Twelve Models. *Journal of Conflict Resolution,* 23: 346–372.

Choucri, N. 1974. *Energy Interdependence.* In *Analyzing Global Interdependence,* Vol. 2. Edited by H. R. Alker, Jr., L. P. Bloomfield, and N. Choucri. Cambridge, Mass.: MIT Center for International Studies.

Choucri, N. (with D. S. Ross and the collaboration of B. Pollins). 1979. *Energy Exchanges and International Relations: A Simulation Model of Petroleum Prices, Trade and Payments.* Cambridge, Mass.: MIT Center for International Studies.

Choucri, N., with V. Ferraro. 1976. *International Politics of Energy Interdependence: The Case of Petroleum.* Lexington, Mass.: Lexington Books. 1976.

Choucri, N. 1980. The International Petroleum Exchange Model: Reference Results and Validation. *Futures,* 12: 201–211.

Choucri, N., D. S. Ross, and D. Meadows. 1976. Toward a Forecasting Model of Energy Politics. *Journal of Peace Science,* 1: 97–111.

Cooper, B. (ed.). 1979. *OPEC Oil Report, Second Edition, November 1979.* London: Petroleum Economist.

Eckaus, R. S. 1973. Absorptive Capacity as a Constraint Due to Maturation Processes. In *Development and Planning: Essays in Honour of Paul Rosenstein-Rodan.* Edited by J. N. Bhagwati and R. S. Eckaus, Cambridge, Mass.: The MIT Press, pp. 79–108.

Eckbo, P. L. 1976. *The Future of World Oil.* Cambridge, Mass.: Ballinger.

Ezzati, A. 1976. Analysis of World Equilibrium Prices, Supply, Demand, Imports and Exports of Crude Oil and Petroleum Products. *Journal of Energy Development,* 1: 306–325.

Federal Energy Administration. 1974. *Project Independence.* Washington, D.C.: Government Printing Office.

Fellner, W. J. 1949. *Competition Among the Few; Oligopoly and Similar Market Structures.* Reprint. New York: A. M. Kelley, 1960.

Fisher, D., D. Gately, and J. F. Kyle. 1975. The Prospect of OPEC: A Critical Survey of Models of the World Oil Market. *Journal of Development Economics,* 2: 363–386.

Ford Energy Policy Project. 1974. *A Time to Choose.* Cambridge, Mass.: Ballinger.

Gersic, J. and G. L. Deyman. 1977. *Factors Affecting World Petroleum Prices to 1985.* Washington, D.C.: International Trade Commission.

Hnyilicza, E. and R. S. Pindyck. 1976. Pricing Policies for a Two-Part Exhaustible Resource Cartel: The Case of OPEC. MITEL 76-008WP. Cambridge, Mass.: MIT Energy Laboratory.

Hotelling, H. 1931. The Economics of Exhaustible Resources. *Journal of Political Economy,* 39: 137–175.

International Monetary Fund. 1979. *International Financial Statistics Yearbook.* Washington, D.C.

Johnston, J. 1972. *Econometric Methods.* New York: McGraw-Hill.

Kalymon, B. A. 1975. Economic Incentives in OPEC Oil Pricing. *Journal of Development Economics*, 2: 337–362.

Kennedy, M. 1974. An Economic Model of the World Oil Market. *Bell Journal of Economics and Management Science*, 4: 540–577.

Kmenta, J. 1971. *Elements of Econometrics*. New York: Macmillan.

Labys, W. C. 1973. *Dynamic Commodity Models: Specification, Estimation and Simulation*. Lexington, Mass.: Lexington Books.

Levy, W. J. 1974. "Implications of World Austerity." Unpublished. New York. Cited in Abolfathi et al. (1977), p. 315.

Moody, J. and Geiger, R. 1975. Petroleum Resources: How Much Oil and Where. *Technology Review*, 77: 38–45.

Murakami, T. 1976. Policy Simulation for Crude Oil Productions of OPEC Countries. *Policy Sciences*, 7: 93–111.

National Petroleum Council. 1965. *Proved Reserves and Productive Capacity of Crude Oil, Natural Gas and Natural Gas Liquids in the United States*. Washington, D.C.: National Petroleum Council.

National Petroleum Council. 1972. *U.S. Energy Outlook*. Washington, D.C.: National Petroleum Council.

New York Times. October 18, 1979, p. D1.

Nordhaus, W. B. 1973. The Allocation of Energy Resources. *Brookings Papers of Economic Activity*, 4: 529–576.

Organization for Economic Cooperation and Development. 1977. *World Energy Outlook*. Paris: OECD.

Organization for Economic Cooperation and Development. 1974. *Energy Prospects to 1985: An Assessment of Long Term Energy Developments and Related Policies*. Vols. 1 and 2. Paris: OECD.

Organization for Economic Cooperation and Development. 1972. *Expenditure Trends in OECD Countries, 1960–1980*. Paris: OECD.

Pindyck, R. S. 1978. Gains to Producers from the Cartelization of Exhaustible Resources. *Review of Economics and Statistics*, 60: 238–251.

Pindyck, R. S. 1978a. OPEC's Threat to the West. *Foreign Policy*, 30: 36–52.

Pugh, A. L. III. 1976. *DYNAMO User's Manual*. 5th ed. Cambridge, Mass.: The MIT Press.

Rybczynski, T. M. 1976. *The Economics of the Oil Crisis*. New York: Holmes and Meir.

Scherer, F. M. 1979. *Industrial Market Structure and Economic Performance*. Chicago: Rand McNally.

Schurr, S. H. and P. T. Homan. 1971. *Middle Eastern Oil and the Western World*. New York: American Elsevier.

Tanzer, M. 1969. *The Political Economy of International Oil and the Underdeveloped Countries.* Boston: Beacon Press.

United Nations, Department of Economic and Social Affairs. 1975. *Demographic Yearbook, 1973.* New York: United Nations.

Workshop on Alternative Energy Strategies. 1977. *Energy: Global Prospects 1985–2000.* New York: McGraw-Hill.

Notes

Notes to Chapter 1

1. These developments are described more fully elsewhere (Choucri 1976).

2. The model is initialized at 1970 values and, to maintain internal consistency, all computations are based on 1970 dollars and then adjusted to 1979 dollars for discussion purposes.

3. The optimal price path is based on the optimization model developed by Pindyck (1978).

4. This falls within the range of recent price estimates by officials of the U.S. Department of Energy and the Central Intelligence Agency. See *The New York Times*, October 18, 1979, p. D1.

Notes to Chapter 2

1. Cartel behavior is one exception. See Fellner (1949) for formal specification, Bobrow and Kudrle (1976) for comparisons with alternative political models, and Choucri (1976) for a preliminary integration of economic and political influences in market interactions.

2. This aggregation is for conceptual clarity and does not impose irrevocable constraints on the results. Nothing in the model precludes a disaggregation of producers, consumers, or companies.

3. In part III sensitivity analysis shows the effect of changing the consumers' reserves and price-responsiveness of non-Gulf producers.

4. The simulation language is DYNAMO, a computer language for modeling complex, nonlinear systems. It recalculates every 0.1 years, with output presented in annual form. See Pugh (1976).

5. This figure is a total estimate, including both recoverable and unrecoverable oil.

6. The non-Gulf suppliers other than the consumers of the West, are the residual suppliers. However, in parts III and IV we model the non-Gulf suppliers explicitly, using the same procedures as the consumers' supply but with different parameters.

7. See Scherer (1979).

8. Min refers to that variable that has the minimum value. Theoretically capacity can exceed reserves. However, production can exceed neither.

9. These variables refer to multiplier functions. In the simulation language used, multipliers are essentially coefficients that can include lag structures.

10. The term CPMOP includes a lag structure.

11. We extrapolate this projection to assume that production increases to 6.5 billion barrels in 1990 and remains level thereafter.

12. The terms CPMOD and COSAM include lag structures.

13. For comparative purposes see the characteristics and results of twelve oil models reviewed in Choucri (1979).

Notes to Chapter 3

1. Adelman (1972), pp. 48–52. Adelman refers to these as "parameters."

2. Although we seek to remain close to Adelman's analysis, some simplification is inevitable. Adelman is the major source of parameter values, interrelationships, and functional specifications.

3. In keeping with Adelman's specifications, we employ the term "proved reserves" synonymously with "recoverable oil-in-place" to refer to the quantity of oil-in-place known to be recoverable given prevailing economic and operating conditions.

4. Adelman (1972), p. 46, argues that it is extremely difficult to provide precise estimates of cost. At the same time, however, he outlines three major components of cost. He differentiates between operating costs (assuming all equipment already on location), development costs (which entail the establishment of new equipment and facilities), and discovery costs (which entail locating new fields and new pools in established fields). For purposes of simplification we differentiate between discovery and development costs in generating a total production cost per barrel and take operating costs as part of the development cost. See note 5 for more recent evidence provided by Adelman on different estimates of production costs.

5. Adelman (1972), p. 47. See also the National Petroleum Council (1965), p. 3.

6. See Adelman (1976).

7. Some additional problems might be pointed out: Adelman (1972, p. 76) states that the development-operating costs in the Gulf during the mid-1960s did not exceed $.10 per barrel. More recently he has argued that development costs have increased by 85 percent since 1969. However, he is not geographically specific (memorandum to Nazli Choucri, June 1977). He also estimates costs in Gulf fields to be about $.25 per barrel. On other occasions he has predicted the production costs in the Gulf area will increase tenfold by 2000 (see Adelman 1976, pp. 8–18). No published explanation of this increase is provided by Adelman.

8. In the absence of more information on the nature of this increase, we stipulated a linear growth over time in 1970 dollars.

9. Consumer import decisions are modeled explicitly elsewhere. The consumer sector of the integrated IPE model computes consumer oil imports, given available substitutes and petroleum from domestic sources.

Notes to Chapter 4

1. Adelman (1972), p. 165.

2. Adelman (1972), p. 209. Adelman notes that with taxes set at $.85 and operating costs at $.05 the companies would receive about $.30 on their investments, a figure that amounts to 26 percent of total gross profit. These estimates are predicated on costs estimated at $.10 and not $.05 per barrel.

3. The fact that oil companies may seek outside financing is a complication not taken into account here.

4. In different sector-simulation runs we investigated the effects of alternative reinvestments by the companies from their oil profits.

5. As presently specified, the producers' oil investments are set at zero; this can be changed to represent changing realities.

Notes to Chapter 5

1. Adelman (1972), p. 82.

2. For detailed and relatively consistent descriptions, see Adelman (1972), Blair (1976), and Tanzer (1969).

3. Adelman (1972) demonstrates this relationship mathematically but does not specify an optimal decline rate. This rate is constrained by the physical nature of the field. Existing documentation suggests that the decline rate in the United States ranges between 8 and 14 percent. In a private correspondence he indicated the optimal rate is about 6.7 percent; the IPE model uses 10 percent to represent a conservative exploration policy.

4. See Blair (1976) for a good review of the issues and Adelman (1972) and Akins (1973) for conflicting views and interpretations of the relationship between oil-rich countries and the international oil companies.

5. It must be stressed that the same can be said for other sectors of the model. In each case information is taken from other sectors and provided in return.

Notes to Chapter 6

1. The end-use price to the consumer would entail an adjustment of these relationships to take into account the consumers' taxes on oil products.

Notes to Chapter 7

1. We have done some sensitivity analyses to determine the effects of the pre-1973 data base. See chapter 11.

2. See Choucri (1979) for a review of twelve oil models.

3. U.N., *Demographic Yearbook, 1973* (1975). Sensitivity analysis can be done with alternative rates of growth.

4. The comments and suggestions of R. S. Eckaus in these formulations are gratefully acknowledged.

5. For a survey of the meanings assigned to absorptive capacity and an interpretation of this concept as a learning and adaptive phenomenon, see R. S. Eckaus, "Absorptive Capacity as a Constraints Due to Maturation Processes," in Bhagwati and Eckaus (1973), pp. 79–108. For a discussion of absorptive capacity and development plans for the oil-exporting countries, see Choucri (1976), pp. 89–114.

6. The estimations were made by D. S. Ross based on Johnston (1972) and Kmenta (1971).

7. These decisions are not separate since a substantial part of imports are for investment purposes.

8. See note 5 regarding absorptive capacity.

9. The fact that, in our model, the distinction among these three variables is represented as an algebraic difference should not obscure the underlying conceptual distinction.

10. The entire process of investment repatriation has been oversimplified due to our assumption that in the short-run repatriation will not occur. A change in this assumption can be accommodated within the equations by modifying the key coefficients and table functions.

Notes to Chapter 8

1. For the most part oil models have adopted an economic rather than an engineering approach to modeling demand. The engineering approach focuses on the petroleum input requirements for specific uses. See the articles in Labys (1973).

2. The same has been said with reference to the oil-exporting countries in chapter 7. So too the relationships can be disaggregated to represent many countries at the same time. The oil-importing countries referred to are the OECD countries, excluding Canada, Australia, and New Zealand.

3. OECD (1974). Volume I presents the general orientation, methods, and findings, and Volume II provides detailed information on procedures and data.

4. The constant 1972 U.S. prices are for Arabian API 34° crude oil f.o.b. at the Persian Gulf.

5. The individual country sources employed by the OECD are listed in *Energy Prospects to 1985*, vol. 1 (1974), p. 43.

6. The OECD study refers to Secretariat work as "Expenditure Trends in OECD Countries 1960–1980" (Paris: OECD, 1972), table B.

7. OECD (1977).

8. Multipliers essentially reflect the effects of elasticities. See chapter 2.

9. Later in this chapter we discuss the differences between the elasticities employed in the IPE model and those in other oil models. It is in relation to these other price elasticities found in the literature that we refer to our own as moderate. See Choucri (1979) for a comparison of the elasticities used in twelve other models of the oil market.

10. This method of determining imports residually is shared by the OECD, vol. 1 (1974), p. 43.

11. Computationally supply from domestic sources is calculated in the same way as demand.

12. Beyond the OECD projections, it is assumed that production rises to 6.5 billion barrels in 1990 and then stabilizes.

13. See OECD, vol. 1 (1974), p. 44.

14. In an earlier version of the IPE model we specified an added, independent effect of technology, taken into account by making provision for a separate technology variable that influences consumer oil production directly. That variable was incorporated to enable us to test different assumptions about the extent of recoverability of reserves. But, since we did not model the process of technological change as such, nor its impact upon oil production in consumer countries, the technology variable then solely allows for the insertion of alternative assumptions regarding the impact of technological change on production and also assures an adequate incorporation of impact.

15. National Petroleum Council (1972).

16. OECD, vol. 2 (1974), p. 11.

17. OECD, vol. 1 (1974), p. 42.

18. In only four studies reviewed in Choucri (1979) are the supply coefficients presented explicitly; a fifth study provides only the supply elasticity for non-OPEC producers. Pindyck (1978) sets the long-run supply elasticity for $12 per barrel at 0.52 and the long-run demand elasticity at 0.90. Kennedy (1974) sets the demand elasticity at 1.0, with no further specification. In no other models are the elasticity coefficients set higher than in these two cases. Ezzati (1976) sets the elasticity of supply for Latin America, Europe, and Asia at 0.1 for 1972 to 1980. In no other model are the elasticity coefficients as low. Kalymon (1975) uses 0.2 as the lower range for the elasticity of supply. Pindyck (1978) gives the most comprehensive and explicit rationale for the short- and long-term elasticities employed for different prices.

19. To examine the effects of different import levels from the Gulf, the fraction would simply need to be modified. One could test different import policies in the same way. See chapter 10 for a respecification of the non-Gulf sources of supply.

20. CIA (1977).

21. This choice of equal weights is the basis for further experimentation and sensitivity analysis and is presented here as an initial point of departure. We do not suggest that the weights are equal. That is essentially an empirical question.

Notes to Chapter 9

1. The balance of payments specification in this chapter does not incorporate gold flows although it could be argued that such inflows are included in the oil-exporting countries' investments in the economies of the importing states, or included in the repatriation of corporate profits. However, the effects may be different in each of these two cases. Including gold flows would not be consistent with the conventional usage of this component in the balance of payments calculations. But, oil-exporting countries' purchases of gold from oil importers have the same effect as oil exporters' investments in oil-importing countries in the balance of payments.

2. In the basic model the producers are specified as having chosen to reinvest all of their investment income, repatriating none of it. In the basic model the producers also chose not to remit any investment income.

3. The variable termed producer investments is initialized at zero in 1970, to determine the model-generated overall investments of the producer countries.

Notes to Chapter 10

1. This is analogous to the distinction made by economists between partial equilibrium and general equilibrium models. Our view of petroleum exchanges is that of general equilibrium which takes into account the interchanges among all related markets.

2. The base is 1970, the initialization year for the model, and 1970 values for the tax rate are inflated by the International Monetary Fund Export Unit Value Index for industrial countries. Results from the reference analysis and all policy analyses are then converted to constant 1979 dollars.

3. The events of 1979 have proved these assumptions invalid. The objective, however, is to utilize some base to examine the internal structure of the model.

4. For the model the OECD projections were extrapolated linearly past 1985.

5. This procedure is designed to represent directionality rather than measure dimensionality. Our computation of markup is somewhat a strange creature, of no theoretical meaning to economists, but one that is full of information.

6. See chapter 11 for a comparison of the results of this alteration with actual production in the 1970s.

Notes to Chapter 11

1. In these comparisons we stress that the IPE model results have not been tuned—this is intentional. The parameters are not adjusted to approximate reality more closely, nor has exogenous tinkering been done to produce better fits. The difficulties of simulation and forecasting in international relations have been widely discussed, and a synthesis of the major issues presented in Choucri (1974) and a comparison of twelve world oil models in Choucri (1979).

2. Although we did not tune the model, a level-base production series is tested here to show the validity of such an assumption; the total results are part of the analysis in part III. Chapter 10 gave the results of some sensitivity analyses.

3. Recall that the OECD demand series vastly *overestimates* demand in comparison with actual data.

4. For supply-related assumptions in these models see Choucri (1979).

Notes to Chapter 12

1. The tax rate corresponding to the optimal price is based on the price per barrel in Pindyck (1978) and computed as follows: (1) discounted values are inflated to give current revenue, and price is then multiplied by demand to yield gross revenue; (2) current revenue is subtracted from gross income to yield total costs; (3) costs are divided by demand to derive the per barrel costs; (4) per barrel costs are subtracted from price, yielding the appropriate tax rate (in 1975 dollars). Since the model is initialized in 1970 dollars, the Pindyck optimal tax rate is deflated using the export unit value price index for the industrialized countries. The results are inflated to 1979 dollars. Pindyck employs an average cost composed of $.50 in 1975 dollars ($.26 in 1970 dollars) growing to just over $1.00 by 2000. See chapter 3 for the IPE model's formulation of cost. Pindyck examines the cartel as a whole; the IPE model is for the Gulf countries only. The parameters and initial values of a simple model of supply and demand used in Pindyck (1978) to generate an optimal price are not transferred into the IPE model. See also Hnyilicza and Pindyck (1976).

2. The peak in 1985 is designed to allow for a juxtaposition with the breakup and the conservative scenarios, rather than for any other political or economic reason.

3. These conclusions, developed with a 5 percent discount rate, are also valid for discount rates of 10 and 20 percent.

Notes to Chapter 13

1. Estimated by Moody and Geiger (1975).

2. From Adelman and Paddock (1979).

3. The supply elasticity in the consumer sector is set at 0.3, see chapter 8. Ezzati (1976) sets the Third World supply elasticity at 0.1.

4. This holds with price elasticities of 0.1 and 0.3.

5. An alternative version of the IPE model disaggregates the producer and consumer countries as well as the non-Gulf and non-Middle East suppliers.

Index